# Recent Advances in Thermoelectric Materials for High Efficiency Energy Conversion and Refrigeration

# Recent Advances in Thermoelectric Materials for High Efficiency Energy Conversion and Refrigeration

Editor

**Paolo Mele**

MDPI • Basel • Beijing • Wuhan • Barcelona • Belgrade • Manchester • Tokyo • Cluj • Tianjin

*Editor*
Paolo Mele
Shibaura Institute of Technology
Japan

*Editorial Office*
MDPI
St. Alban-Anlage 66
4052 Basel, Switzerland

This is a reprint of articles from the Special Issue published online in the open access journal *Materials* (ISSN 1996-1944) (available at: https://www.mdpi.com/journal/materials/special_issues/thermoelectr_mater).

For citation purposes, cite each article independently as indicated on the article page online and as indicated below:

LastName, A.A.; LastName, B.B.; LastName, C.C. Article Title. *Journal Name* **Year**, *Volume Number*, Page Range.

**ISBN 978-3-0365-3503-6 (Hbk)**
**ISBN 978-3-0365-3504-3 (PDF)**

# Contents

# About the Editor

**Paolo Mele** is currently Professor at SIT Research Laboratories, Shibaura Institute of Technology, Tokyo, Japan. He obtained a Master's degree in Chemistry and a Ph.D. in Chemical Sciences at Genova University (Italy). In 2003, he moved to ISTEC-SRL in Tokyo to study melt-textured ceramic superconductors. Then, he worked as a postdoctoral researcher at Kyoto University (JSPS fellowship) from 2004 to 2007, at the Kyushu Institute of Technology (JST fellowship) from 2007 to 2011, at Hiroshima University (as a lecturer) from 2011 to 2014, and at Muroran Institute of Technology (as an associate professor) from 2015 to 2018 before taking up his current position. His research interests include materials for energy and sustainable development (superconductors and thermoelectrics); the fabrication and characterization of thin films of oxides, ceramics, and metals; the study of the effect of nanostructuration on the physical properties; thermal transport; and vortex matter. He is the author of more than 110 papers in international scientific journals and four book chapters, has two patents, and has contributed to hundreds of communications at international conferences. He has co-edited nine books on superconductors, oxides, thin films, thermoelectrics, and other materials for energy.

MDPI

*Editorial*

# Special Issue "Recent Advances in Thermoelectric Materials for High Efficiency Energy Conversion and Refrigeration"

**Paolo Mele** [1,2]

[1] College of Engineering, Innovative Global Program, Shibaura Institute of Technology, 307 Fukasaku, Minuma-ku, Saitama 337-8570, Japan; pmele@shibaura-it.ac.jp
[2] International Center for Green Electronics, Shibaura Institute of Technology, 3-7-5 Toyosu, Koto-ku, Tokyo 135-8548, Japan

**Citation:** Mele, P. Special Issue "Recent Advances in Thermoelectric Materials for High Efficiency Energy Conversion and Refrigeration". *Materials* **2022**, *15*, 1672. https://doi.org/10.3390/ma15051672

Received: 5 February 2022
Accepted: 16 February 2022
Published: 23 February 2022

**Publisher's Note:** MDPI stays neutral with regard to jurisdictional claims in published maps and institutional affiliations.

It has been almost three years since I enthusiastically accepted to be guest editor for this Special Issue of *Materials*, entitled "Recent Advances in Thermoelectric Materials for High Efficiency Energy Conversion and Refrigeration".

Thermoelectricity is a well-known phenomenon enabling the conversion of heat into electric energy without moving parts. Its exploitation has been widely considered to contribute to the increasing need for energy along with the concerns about the environmental impact of traditional fossil energy sources. In the last few years, significant improvements in the performance of thermoelectric materials have been achieved through chemical doping, solid solution formation, and nanoengineering approaches. Furthermore, the feasibility of flexible, stretchable, and conformable thermoelectric harvesters has been demonstrated and has attracted the interest of a wide audience. However, the path for practical applications of thermoelectrics still appears long.

This Special Issue of *Materials* was intended as an effort to bridge the gap between materials science and applications of thermoelectric materials.

Originally, many topics were welcome: new thermoelectric compounds; correlation between material structure and thermoelectric properties; bulk thermoelectric ceramics, oxides, and chalcogenides; bulk thermoelectric alloys and intermetallics; organic and polymeric thermoelectrics; thermoelectric thin films, multilayers, and nanocomposites; theory and modeling; thermal transport and thermal conductivity; applications and devices based on thermoelectric materials; standardization and metrology; and more.

The five published papers reflect the original spirit of the Special Issue, going from skutterudite materials to sulfides and oxides and covering various aspects of materials preparation and characterization.

With the great hope that it will contribute to pave the way for wide diffusion of thermoelectric materials in society and daily life, I declare this Special Issue closed thanking all the colleagues and all the editorial staff of *Materials* for their great contributions and everlasting support.

**Funding:** This research received no external funding.

**Conflicts of Interest:** The author declares no conflict of interest.

 *materials*

*Article*

# Redox-Promoted Tailoring of the High-Temperature Electrical Performance in $Ca_3Co_4O_9$ Thermoelectric Materials by Metallic Cobalt Addition

Gabriel Constantinescu *, Artur R. Sarabando, Shahed Rasekh, Diogo Lopes, Sergii Sergiienko, Parisa Amirkhizi, Jorge R. Frade and Andrei V. Kovalevsky

Department of Materials and Ceramic Engineering, CICECO–Aveiro Institute of Materials, University of Aveiro, 3810-193 Aveiro, Portugal; artursarabando@ua.pt (A.R.S.); shahedvrm@ua.pt (S.R.); djlopes@ua.pt (D.L.); sergiienko@ua.pt (S.S.); parisa.amirkhizi@ua.pt (P.A.); jfrade@ua.pt (J.R.F.); akavaleuski@ua.pt (A.V.K.)
* Correspondence: gabriel.constantinescu@ua.pt

Received: 27 January 2020; Accepted: 25 February 2020; Published: 27 February 2020

**Abstract:** This paper reports a novel composite-based processing route for improving the electrical performance of $Ca_3Co_4O_9$ thermoelectric (TE) ceramics. The approach involves the addition of metallic Co, acting as a pore filler on oxidation, and considers two simple sintering schemes. The $(1-x)Ca_3Co_4O_9/xCo$ composites (x = 0%, 3%, 6% and 9% vol.) have been prepared through a modified Pechini method, followed by one- and two-stage sintering, to produce low-density (one-stage, 1ST) and high-density (two-stage, 2ST) ceramic samples. Their high-temperature TE properties, namely the electrical conductivity ($\sigma$), Seebeck coefficient ($\alpha$) and power factor (PF), were investigated between 475 and 975 K, in air flow, and related to their respective phase composition, morphology and microstructure. For the 1ST case, the porous samples (56%–61% of $\rho$th) reached maximum PF values of around 210 and 140 $\mu Wm^{-1}\cdot K^{-2}$ for the 3% and 6% vol. Co-added samples, respectively, being around two and 1.3 times higher than those of the pure $Ca_3Co_4O_9$ matrix. Although 2ST sintering resulted in rather dense samples (80% of $\rho$th), the efficiency of the proposed approach, in this case, was limited by the complex phase composition of the corresponding ceramics, impeding the electronic transport and resulting in an electrical performance below that measured for the $Ca_3Co_4O_9$ matrix (224 $\mu Wm^{-1}\cdot K^{-2}$ at 975K).

**Keywords:** calcium cobaltite; TE performance; electrical properties; composite; redox tuning

---

## 1. Introduction

Thermoelectric (TE) materials can directly convert an applied temperature gradient into electrical voltage due to the Seebeck effect and are regarded as a promising solution for producing electrical power from waste-heat sources [1–3]. They are employed in self-sufficient, robust TE devices (modules and generators), which are very reliable, sustainable and scalable, allowing mainly for mobile or remote applications [4,5]. The range of possible applications for TE materials is mostly limited by their relatively low conversion efficiencies [6], but with the recent aid of machine learning and artificial intelligence tools, new horizons in TE materials are envisaged [7]. The performance of a TE material is limited by the Carnot efficiency and is quantified through the dimensionless figure of merit ZT:

$$ZT = \frac{\alpha^2 \sigma}{\kappa} T \tag{1}$$

combining the absolute Seebeck coefficient ($\alpha$), electrical conductivity ($\sigma$), total thermal conductivity ($\kappa$) and prospective working temperature (T). The electrical part of ZT ($\alpha^2\sigma$) is called the power factor (PF) and depends entirely on the material's intrinsic electrical properties. It becomes obvious from the expression of ZT that good TE materials must simultaneously possess large $\alpha$ and $\sigma$ and small $\kappa$. These TE coefficients, however, are not independent of each other and cannot be treated separately, without affecting the others. For example, the Wiedemann–Franz law addresses the intimate fundamental relationship between the electrical conductivity and the electronic contribution to $\kappa$. Therefore, the usual approaches in improving ZT [8] are the decrease of phonon contribution to $\kappa$ and/or the increase in PF. As most established TE materials are semiconductors, the PF presents a maximum value within a narrow range of charge carrier concentration. The optimization of carrier concentration is usually performed through various electronic band-structure engineering techniques [9–11].

Established TE materials like $Bi_2Te_3$, $Bi_2Se_3$, PbTe, half-Heusler alloys, intermetallic Zintl phases, skutterudite and some Si-based alloys have already demonstrated feasible power generation performances (ZT $\approx$ 1) at low and intermediate temperature ranges [12,13]. At temperatures above ~800–900 K in air, however, they do not possess the necessary thermal and/or chemical stability needed for power generation applications, and they degrade or decompose. Furthermore, they contain expensive, toxic and/or scarce elements which impose important limitations. For these reasons, the established TE materials are usually employed in niche situations where their advantages outweigh their disadvantages [14].

With the discovery of attractive TE properties in $Na_xCoO_2$ ceramics in 1997 [15], a lot of effort has been put in the research and development of CoO-based materials, as well as other transition metal oxides [16], which have important 'default' advantages (abundance, low-cost, environmental 'friendliness', low reactivity and high thermochemical stability) over established TE materials, enabling them to be considered for power generation applications at high temperatures and in oxidizing conditions [17–21]. While the best performing n-type TE oxides were found in the family of perovskite-type titanates [22–25], manganites [26–29] and ZnO-based materials [30–32], one of the most promising p-type TE materials (considered as the best choice for a p-type leg in a high-temperature TE module) continues to be the so-called $Ca_3Co_4O_9$ compound, belonging to the family of misfit-layered cobaltites [26]. However, its main practical drawbacks, namely the strong anisotropic electrical properties induced by the particular misfit crystal structure (which promotes the growth of elongated, randomly oriented plate-like grains of different shapes and sizes, resulting in space-inefficient packing in the bulk ceramics and poor electrical contacts between the grains, under classical sintering/consolidation methods/conditions) and the low bulk density and weak mechanical strength (due to the big difference between the maximum stability temperature of the $Ca_3Co_4O_9$ phase and the temperature of its liquid phase), represent some of the major challenges in the development of TE modules using $Ca_3Co_4O_9$ as a p-type leg.

The $Ca_3Co_4O_9$ compound is an intrinsically nanostructured material that has a monoclinic crystal structure (P2(3) space group) consisting of two different alternating layers, stacked in the $c$-axis direction, namely a distorted triple rock-salt (RS) type $Ca_2CoO_3$ insulating layer, where the cobalt valence is (2+), which is sandwiched between two hexagonal (H) $CdI_2$-type $CoO_2$ conductive layers (consisting of edge-shared $CoO_6$ octahedra), where the mean cobalt valence is between (3+) and (4+). A complex misfit structure is built along the $b$-axis direction [33,34], since the $b$ cell parameters are incommensurate (the $a$- and $c$-axes and $\beta$ angles are common). It is worth mentioning that the $CoO_2$ crystallographic planes from $Ca_3Co_4O_9$ are isostructural to those found in the superconductor $Na_{0.35}CoO_2 \times 1.3H_2O$, with $T_c$ = ~5 K [35]. Based on its crystal structure, $Ca_3Co_4O_9$ is more technically written with the chemical formula $[Ca_2CoO_3][CoO_2]_{1.62}$, where '1.62' is the so-called incommensurability ratio ($b_{RS}/b_H$), found to be responsible for the high Seebeck coefficient values [36], which are susceptible to modifications using chemical substitutions.

Some of the best-known individual or combined approaches for enhancing the TE performances of $Ca_3Co_4O_9$ include cation substitutions in both calcium and cobalt sites, microstructural engineering techniques and some composite approaches [37–45]. The density and grain connectivity may be improved by using specific processing methods, like SPS and LFZ [46–48], and high-quality, ultrafine precursor powders [49–51].

Two-step sintering schemes also produce quite dense ceramics, but they require long processing/annealing times to stabilize the TE phase [48] and to avoid the formation of unwanted additional secondary phases. According to the equilibrium phase diagram of the Ca-Co-O system in air [52,53], the $Ca_3Co_4O_9$ phase is stable up to 1199 K, after this limit decomposing to $Ca_3Co_2O_6$ and CoO, which are both stable up to 1299 K, following the decomposition reactions from the following equations:

$$Ca_3Co_4O_9 \rightarrow Ca_3Co_2O_6 + 2\,CoO + O_2 \uparrow \tag{2}$$

$$Ca_3Co_2O_6 \rightarrow 3\,CaO\,(ss) + 2\,CoO\,(ss) + O_2 \uparrow \tag{3}$$

Following the stoichiometry line further, the first liquid phase appears at 1623 K, a fact which is very important when considering problems related to the bulk density of this phase.

This work aims to study a redox-promoted approach for improving the electrical performance of the bulk $Ca_3Co_4O_9$ compound, implying an addition of a dispersed metallic phase which further oxidizes on sintering and provides a pore-filling effect. The efficacy of the proposed approach is assessed by measuring the high-temperature TE properties of the resulting composite materials, which are further related to their compositions, morphologies and microstructures.

## 2. Materials and Methods

The $Ca_3Co_4O_9$ ceramic matrix materials were prepared from ultrafine $Ca_3Co_4O_9$ precursor powders (batches of ~10 g), obtained through a modified Pechini method. Micrometric metallic Co powder (Alfa Aesar, 1.6 μm, 99.8%, metals basis) was weighed in the stoichiometric amount and dissolved in a minimum amount of medium concentrated nitric acid ($HNO_3$, LabKem 65% AGR ISO, ACS) and distilled water. The mixture was slowly heated and magnetically stirred (on a hotplate, with a magnetic stirrer), to evaporate the excess water and to produce the cobalt nitrate, until a relatively viscous pink/violet gel formed. Distilled water and the stoichiometric amount of citric acid monohydrate (Sigma-Aldrich, ACS reagent, ≥99.0%) was added afterward to the mixture, while slowly stirring and heating it continuously to around 423 K, for a minimum of 2 h, and adding distilled water from time to time, to maintain a minimum volume of liquid in the beaker. After this step, the stoichiometric amount of $CaCO_3$ powder (Sigma-Aldrich, ACS reagent, chelometric standard, 99.95–100.05% dry basis) was added and left together to mix in the same conditions for ~30 min. Finally, the necessary measured volume of ethylene glycol (Fluka Analytical, puriss. p.a., Reag. Ph. Eur., ≥99.5%) was added to the pink/violet liquid, which started the polymerization reaction, signaled by the appearance of bubbles which later turned to foam, aided by controlling the optimal temperature and stirring values. The liquid mixture from this last part of the modified Pechini chemical synthesis process (which lasted from 3 to 5 h) was subjected to temperatures between 473 and 573 K, on the hotplate; this contributed to the chemical reactions to occur, changing in the last part, after the evaporation of distilled water excess, to a very viscous pink/violet gel or paste, which eventually solidified completely. At this step, the stirring was stopped completely, and the burning of the organic components began, using temperatures between 573 and 673 K, and times between 2 and 4 h. The resulted black powder was finely ground in an agate mortar (aided by high-purity ethanol, which was evaporated afterward) and subjected to a 3-step thermal treatment cycle in air (room temperature (RT) to 573 K (5 K/min), to 873 K (1 K/min, dwell 6 h), to 1073 K (1 K/min, dwell 6 h), and to RT (5 K/min)), in order to decompose the carbonates and burn-out the organic phases excess (controlled combustion), promoting the formation of the desired TE $Ca_3Co_4O_9$ phase.

The calcined precursor powders were finely ground one last time, in the same way as before, and the various compositions have been produced/prepared, by adding 3%, 6% and 9% vol. of metallic Co and mixing them together, thoroughly. Pure matrix compositions were kept for comparison, as references. The final compositions were uniaxially pressed at 200 MPa (15.7 kN) for around one minute. The green ceramic pellets were subsequently sintered in air, following 2 different sintering schemes inspired from References [49] and [54], to produce low-density and high-density samples, respectively:

One-stage, 1ST: RT to 1173 K (2 K/min, dwell 24 h), to RT (2 K/min).

Two-stage, 2ST: RT to 773 K (8 K/min), to 1473 K (2 K/min, dwell 6 h), to 1173 K (10 K/min, dwell 72 h), to RT (2 K/min).

After sintering, the pellets were carefully polished, finely ground or cut in the adequate shapes and sizes, for the relevant characterization to be performed onward. The experimental densities ($\rho_{exp}$) of $Ca_3Co_4O_9$-based ceramic samples were determined by geometrical measurements and weighing. The estimated errors in all cases were found to be <3%.

Phase identification was performed through powder X-Ray Diffraction (XRD) analyses, for various $Ca_3Co_4O_9$-based samples (ground into powder) and for precursors (after the organic phases burn-out and after the 3-step thermal treatment), at RT, using a PHILIPS X'PERT system with $CuK_\alpha$ radiation ($Cu_\alpha$ = 1.54060 Å), with 2θ angles ranging between 5 and 90 degrees and a step and exposure time of 0.02°2θ and 3 s, respectively.

Morphological characterization of fractured samples coated with carbon was performed using scanning electron microscopy (SEM, Hitachi SU-70 instrument, Aveiro, Portugal), complemented by energy-dispersive spectrometry (EDS, Bruker Quantax 400 detector).

Electrical conductivity and Seebeck coefficient measurements were simultaneously performed on bar-shaped samples ($\sim$10 × 2 × 2 mm), in constant air flow, using a custom setup described in detail elsewhere [55], from 475 to 975 K, with a step of 50 K, using a steady-state technique. Freshly cut samples of each composition were fixed inside a specially designed alumina sample holder, placed in turn inside a high-temperature furnace, one horizontally ($\sigma$, electrically connected with fine Pt wire, following a four-point probe DC technique arrangement, using an applied electric current) and the other vertically ($\alpha$, subjected to a local constant temperature difference of $\sim$14 K). The measurement part of the custom setup was similar to that described in Reference [56]. From the measured $\sigma$ and $\alpha$ values, the PF values were calculated in each case, at each temperature step.

The XRD and TE coefficient plots were constructed using the OrginPro software (2019b (9.65)).

## 3. Results and Discussions

### 3.1. Structural Characterization

The approach proposed in the present work involves the redox-promoted tailoring of the microstructural features, which are known to be of particular importance for the TE performance of $Ca_3Co_4O_9$. To avoid the formation of excessively porous material with inhomogeneous cations distribution, a chemical synthesis route based on combustion (modified Pechini) was chosen, which provides the necessary high-quality precursor powder, possessing high reactivity, homogeneity and low particle size, leading to desired, single-phase compositions in the case of reference samples [49].

From the XRD pattern for dried precursor powder sample, taken immediately after the organic phases combustion/decomposition/burn-out ('dried precursor', Figure 1A), one can clearly see that the present phases are mainly calcium carbonate and cobalt oxides, in agreement with previously reported results for a similar case [49]. The burn-out temperature ($\sim$623 K) from this instance is not high enough to form the desired TE phase.

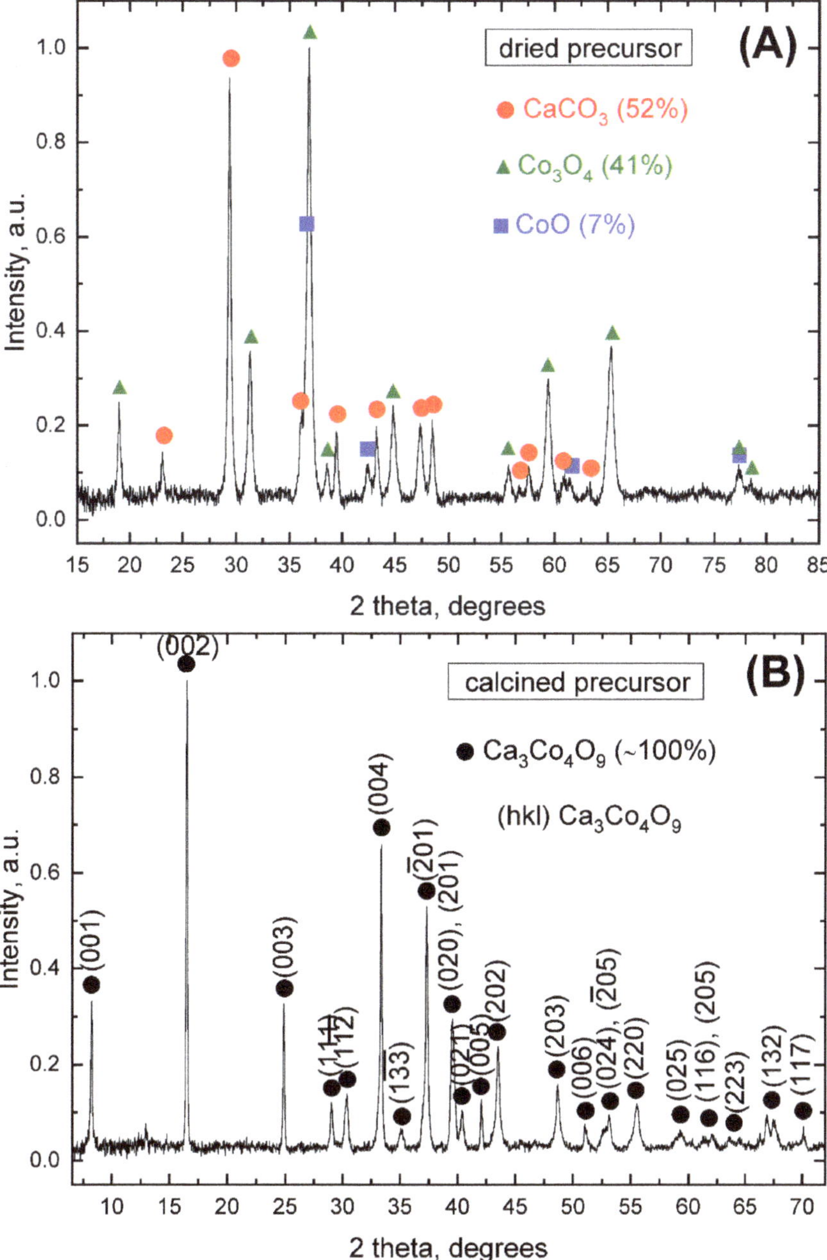

**Figure 1.** Normalized XRD patterns of the modified Pechini precursor powder: (**A**) after drying at ~623 K for ~3 h (organic phases burn-out) and (**B**) after the three-step thermal treatment at 573 K, 873 K (6 h) and 1023 K (6 h), showing the phase composition and estimated amount in each case. The (hkl) crystallographic planes belonging to the $Ca_3Co_4O_9$ phase are also shown in (**B**).

In agreement with the corresponding equilibrium phase diagram [52,53], the formation of the $Ca_3Co_4O_9$ phase starts taking place after the application of the three-step annealing cycle, which promotes the following reactions to occur:

$$6CoO + O_2 \rightarrow 2Co_3O_4 \tag{4}$$

$$9CaCO_3 + 4Co_3O_4 + O_2 \rightarrow 3Ca_3Co_4O_9 + 9\,CO_2 \uparrow \tag{5}$$

resulting in nearly single-phase, high-quality precursor powder. The respective XRD pattern can be seen in Figure 1B ('calcined precursor'), where the corresponding peaks and (hkl) crystallographic planes belong to the $Ca_3Co_4O_9$ phase, as shown by the PDF cards #00-058-0661 [33] and #00-062-0692.

Going forward to the 1ST and 2ST sintered samples obtained from the modified Pechini precursor powder, the XRD results (Figure 2A–F) clearly show the presence of additional factors affecting the final phase composition in all cases, being more simple in the 1ST case and more complex in the 2ST case, as compared to the pure matrix samples. From now on, the 0%, 3%, 6% and 9% vol. Co containing $Ca_3Co_4O_9$ samples sintered in one and two stages will be denoted as 0, 3, 6 or 9Co_1ST or 2ST, for simplicity, as shown in Table 1. Table 1 also shows the fractions of $Ca_3Co_4O_9$ phase and phase impurities in wt.%, as estimated by the RIR method.

**Table 1.** Denominations, phase composition and density of the prepared ceramic samples.

| Composition | Processing Conditions | Denomination | Phase Composition, wt.%* | Density $\rho_{exp}$, g/cm$^3$ | $\rho_{exp}/\rho_{th}$** |
|---|---|---|---|---|---|
| $Ca_3Co_4O_9$ | one-stage | 0Co_1ST | $Ca_3Co_4O_9$(100) | 2.62 | 0.56 |
| $Ca_3Co_4O_9$ +3% vol. Co | one-stage | 3Co_1ST | $Ca_3Co_4O_9$(94); $Co_3O_4$(6) | 2.90 | 0.61 |
| $Ca_3Co_4O_9$ + 6% vol. Co | one-stage | 6Co_1ST | $Ca_3Co_4O_9$(85); $Co_3O_4$(15) | 2.85 | 0.59 |
| $Ca_3Co_4O_9$ + 9% vol. Co | one-stage | 9Co_1ST | $Ca_3Co_4O_9$(80); $Co_3O_4$(20) | 2.81 | 0.57 |
| $Ca_3Co_4O_9$ | two-stage | 0Co_2ST | $Ca_3Co_4O_9$(100) | 3.74 | 0.80 |
| $Ca_3Co_4O_9$ + 3% vol. Co | two-stage | 3Co_2ST | $Ca_3Co_4O_9$(94); $Ca_3Co_2O_6$(4); $Co_3O_4$(2) | 4.12 | - |
| $Ca_3Co_4O_9$ + 6% vol. Co | two-stage | 6Co_2ST | $Ca_3Co_4O_9$(70); $Ca_3Co_2O_6$(23); $Co_3O_4$(7) | 4.35 | - |
| $Ca_3Co_4O_9$ + 9% vol. Co | two-stage | 9Co_2ST | $Ca_3Co_4O_9$(40); $Ca_3Co_2O_6$(40); $Co_3O_4$(20) | 4.49 | - |

*—Estimated using the RIR method; **—Theoretical density.

Firstly, the XRD patterns corresponding to both pure matrix compositions 0Co_1ST and 0Co_2ST (Figure 2A,B) indicate the presence of single-phase $Ca_3Co_4O_9$, as marked by the corresponding (hkl) crystal planes, in agreement with the work of Masset et al. [33] and other literature references [52,53]. Thus, in terms of the phase composition, there is no significant difference between the two sintering schemes applied to the pure matrix samples; these samples are further used as references to follow the effects imposed by the cobalt additions. The XRD data for the Co-containing samples (3, 6 and 9Co_1ST and 2ST) suggest the presence of new phases, different for each sintering scheme, as shown in Figure 2C–F. Those secondary phases correspond to $Co_3O_4$ for the 1ST case, and $Co_3O_4$ and $Ca_3Co_2O_6$ for the 2ST sintered samples, respectively. Their concentration, estimated using the RIR method, increases with the addition of metallic Co (Table 1).

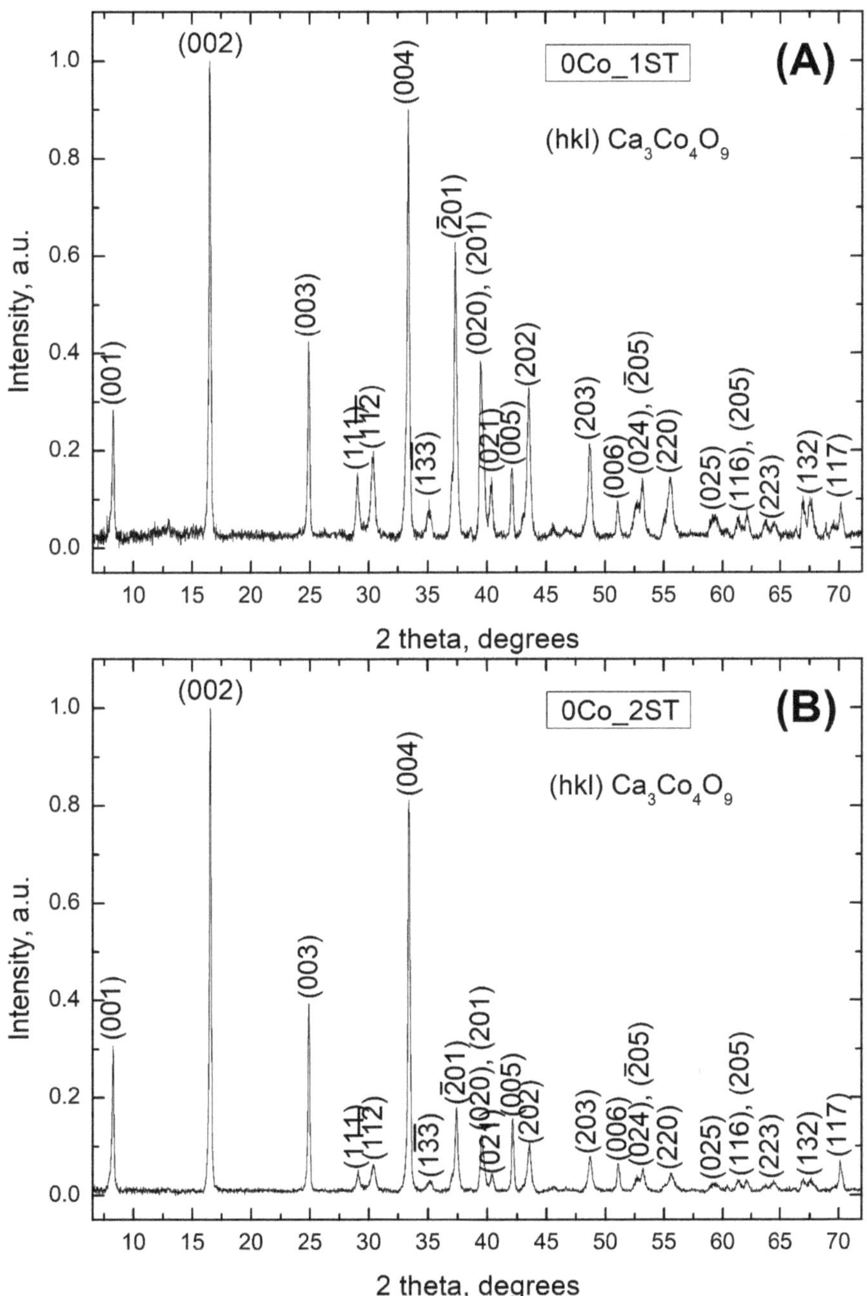



**Figure 2.** *Cont.*

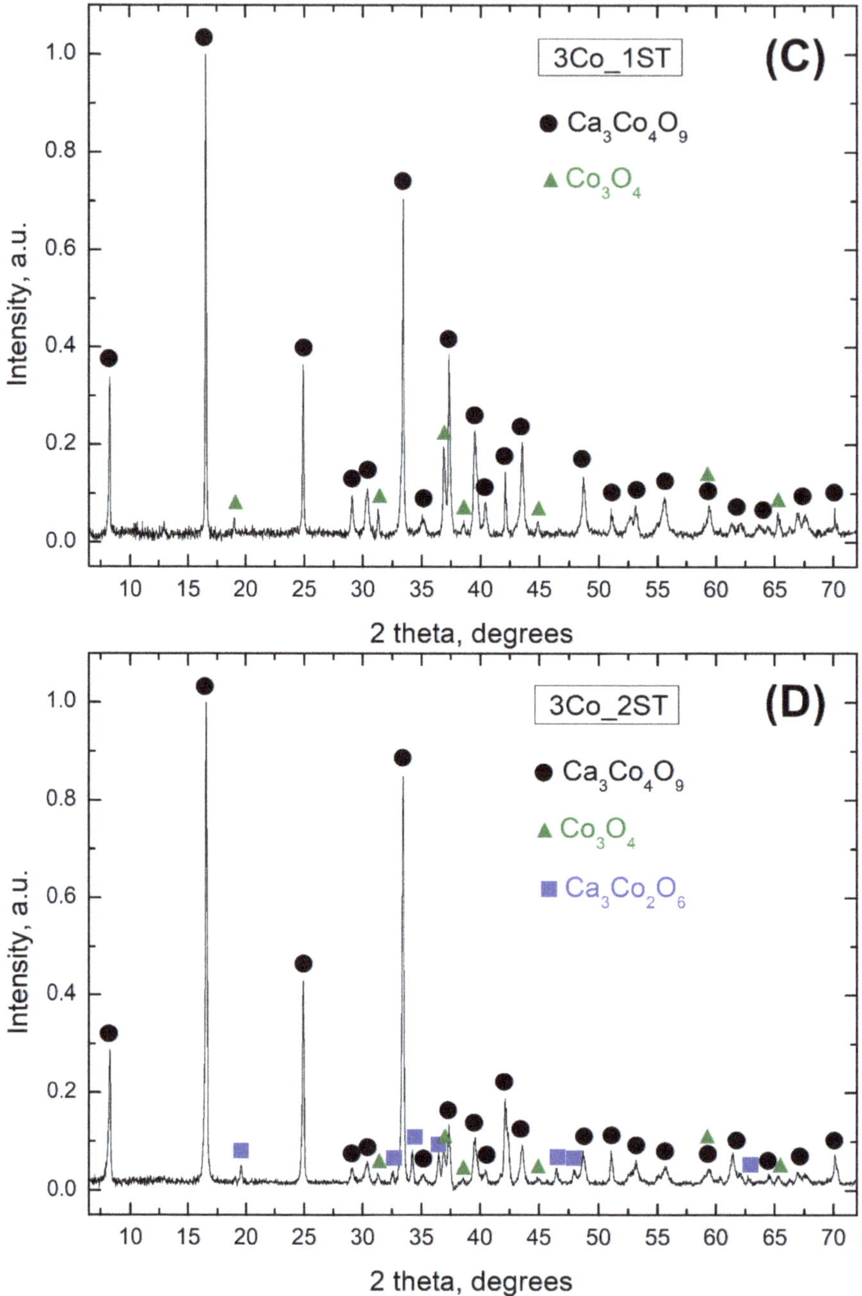


**Figure 2.** *Cont.*

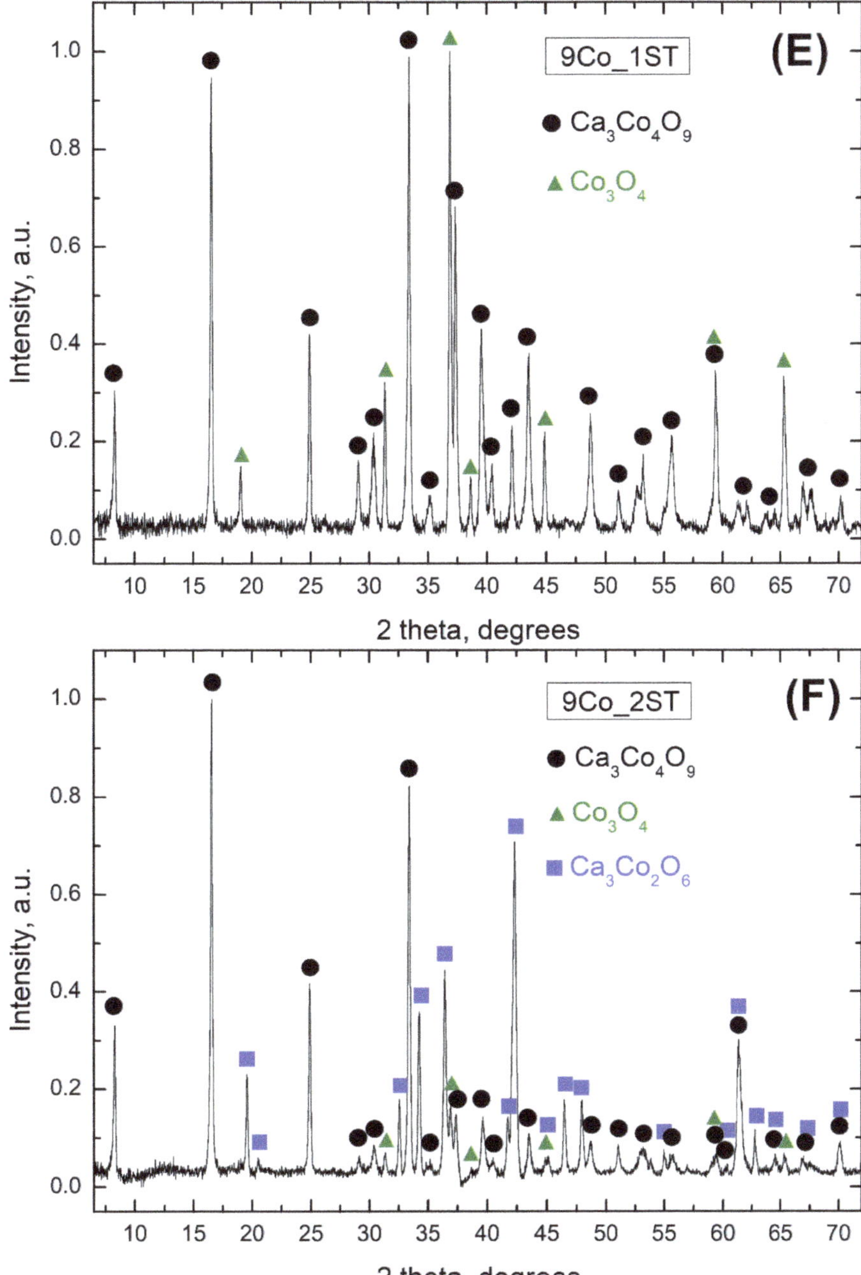

**Figure 2.** Normalized XRD patterns of the 1ST and 2ST sintered samples: (**A**) 0Co_1ST, (**B**) 0Co_2ST, (**C**) 3Co_1ST, (**D**) 3Co_2ST, (**E**) 9Co_1ST and (**F**) 9Co_2ST, showing the phase composition in each case. In (**A,B**), the (hkl) crystallographic planes belonging to the $Ca_3Co_4O_9$ phase are shown.

Metallic cobalt readily oxidizes in the air, at temperatures above 900 K, forming a CoO and $Co_3O_4$ mixture [57]. Hence, the observed phase composition is in good agreement with that predicted by the phase diagrams [52,53]. Shifting from the nominal $3CaO$–$2Co_2O_3$ ratio to the Co-rich region mainly promotes the formation of $Ca_3Co_4O_9$ + $Co_3O_4$ at temperatures below 1060–1200 K (1ST case), while the formation of $Ca_3Co_2O_6$ and $Co(Ca)O$ takes place at higher temperatures (2ST case). The re-equilibration of the $Ca_3Co_2O_6$ and $Co(Ca)O$ mixture at 1173 K during the second step from the 2ST sintering is only partial, explaining why the resulting composites from this case still retain a significant fraction of $Ca_3Co_2O_6$ phase.

### 3.2. Microstructural Evolution

Some additional insights on the observed difference in phase composition between the 1ST and 2ST sintered samples can be obtained by analyzing the densities of the prepared ceramic materials (Table 1). The relative density of the 1ST samples was calculated by assuming a simple mixing rule and the theoretical densities of 4.69 and 6.06 $g/cm^3$ for $Ca_3Co_4O_9$ and $Co_3O_4$, correspondingly [58]. The density shows a noticeable improvement for the 3Co_1ST samples, as compared to the pure matrix; the values for 6Co_1ST and 9Co_1ST are also slightly higher than for the reference, but slightly lower than for 3Co_1ST. Together with the electron microscopy results discussed below, these results suggest that the addition of metallic cobalt particles contributes by filling the undue porosity during oxidation. This porosity is formed by the plate-like grains of different shapes and sizes and by the low packing density resulting from the 1ST sintering scheme. At high cobalt contents, this effect vanishes, probably due to the excessive expansion provided by the cobalt oxide formation and corresponding detachment of the particles/grains. It should be noticed that the 9Co_1ST samples also show a relatively poor mechanical strength, as compared to the other 1ST samples. In any case, the apparently low-density values of the 1ST sintered samples are typical for this material, obtained in this manner [59].

On the other hand, rather dense samples were obtained in the 2ST sintering case. The relative density was not assessed for these samples due to possible errors originating from a more complex phase composition. Nevertheless, the tendency for a general density increase as compared to the pure matrix composition is evident, the corresponding $\rho_{exp}$ being comparable with those found in the literature, e.g., for microwave processed samples [40]. Although simultaneously resulting in a more complex phase composition, the relatively high processing temperature actually facilitates the densification of the samples. At the same time, the first temperature step at 1473 K significantly suppresses the oxygen exchange/diffusion in the samples and delays the phase composition equilibration during the second step, 1173 K, in the window where the TE $Ca_3Co_4O_9$ phase is formed. Thus, the final composition of the 2ST sintered samples correspond to a kinetically frozen state, representing a significant fraction of high-temperature $Ca_3Co_2O_6$ phase. Similar observations have been reported elsewhere [48].

The significant porosity of the 1ST sintered samples, known to be detrimental for the electrical performance, actually facilitates the thermal equilibration, due to a faster oxygen diffusion, which results in the formation of a higher amount of $Ca_3Co_4O_9$ phase for the same nominal composition, as compared to the equivalent 2ST case.

The above discussion is in good agreement with the following microstructural results (Figures 3 and 4). For the 1ST case, the representative SEM images from fractured samples shown in Figure 3A,C,E apparently suggest an improvement in densification from 0Co_1ST to 3Co_1ST, while the morphology of the grains remains essentially unchanged. The high porosity found for the 9Co_1ST sample is also visible in the micrograph E (Figure 3). The small grain sizes and low particle-size dispersion, typical for the Pechini precursor powders, can also be seen in these selected SEM pictures from the 1ST sintered samples, where the mean grain sizes are estimated to be ~1 μm.

For the 2ST case, however, the morphologies and microstructures of the samples change substantially, especially for the Co-containing ones (Figure 3B,D,F). First of all, the porosity is much lower than in the 1ST case, promoted by the additional high-temperature sintering step, which, together with the long annealing time, also led to a twofold or threefold increase in grain size (Figure 3B) and a better consolidation during sintering. Furthermore, when Co is added to the $Ca_3Co_4O_9$ matrix, the plate-like grains associated to the $Ca_3Co_4O_9$ phase become increasingly harder to identify in the micrographs of the corresponding samples, due to the increase in secondary phases amount (especially $Ca_3Co_2O_6$), in agreement with the corresponding XRD results shown previously.

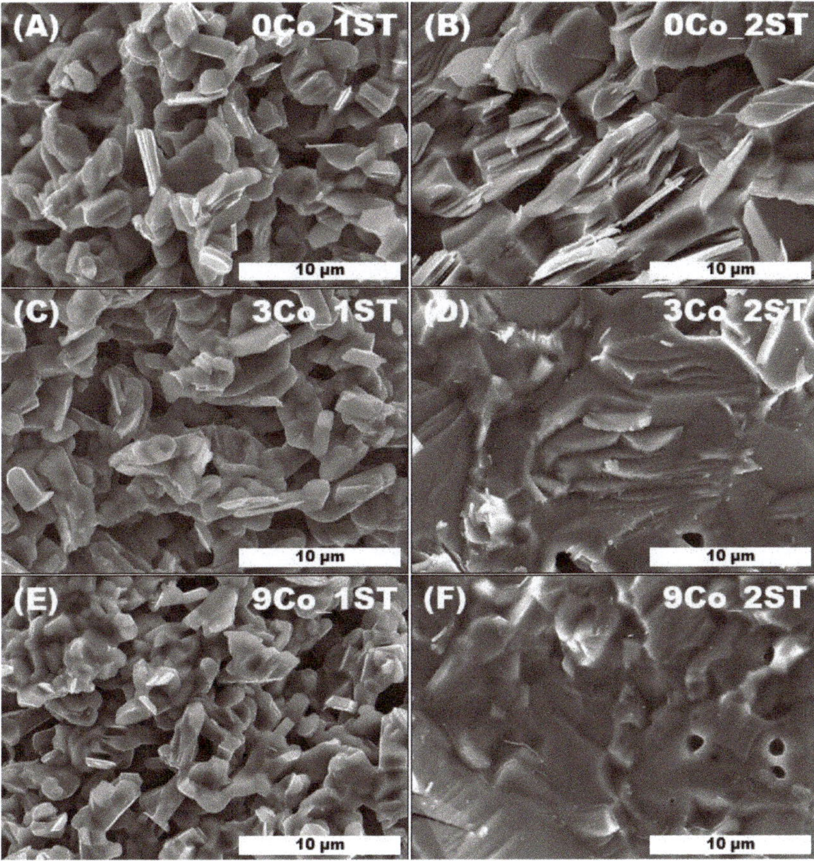

**Figure 3.** Representative SEM micrographs of fractured 1ST and 2ST sintered samples: (**A**) 0Co_1ST, (**B**) 0Co_2ST, (**C**) 3Co_1ST, (**D**) 3Co_2ST, (**E**) 9Co_1ST and (**F**) 9Co_2ST.

The respective representative EDS maps from Figure 4 show more details on the phase evolution and explain the different morphologies and microstructures found for the two sintering schemes employed. Firstly, the EDS maps from Figure 4A,B suggest a uniform Ca and Co distribution in the 0Co_1ST and 0CO_2ST samples, in agreement with their pure phase composition (Figure 2A,B and Table 1), showing 100% $Ca_3Co_4O_9$. The microstructural arrangements observed for 3Co_1ST and, especially, 9Co_1ST (Figure 4C,E) indicate that $Co_3O_4$, which forms on oxidation of Co particles simultaneously with the formation of $Ca_3Co_4O_9$, may actually connect highly asymmetric $Ca_3Co_4O_9$ grains between themselves, acting as a pore filler in the 1ST samples with low-density grain packing. These connections are fewer and more homogeneously distributed for 3Co_1ST, while their number increases and their homogeneity decreases (agglomerates start to form gradually) on increasing the amount of added cobalt. The large agglomerates of $Co_3O_4$ from 9Co_1ST may explain the higher porosity and subsequent lower density of these samples.

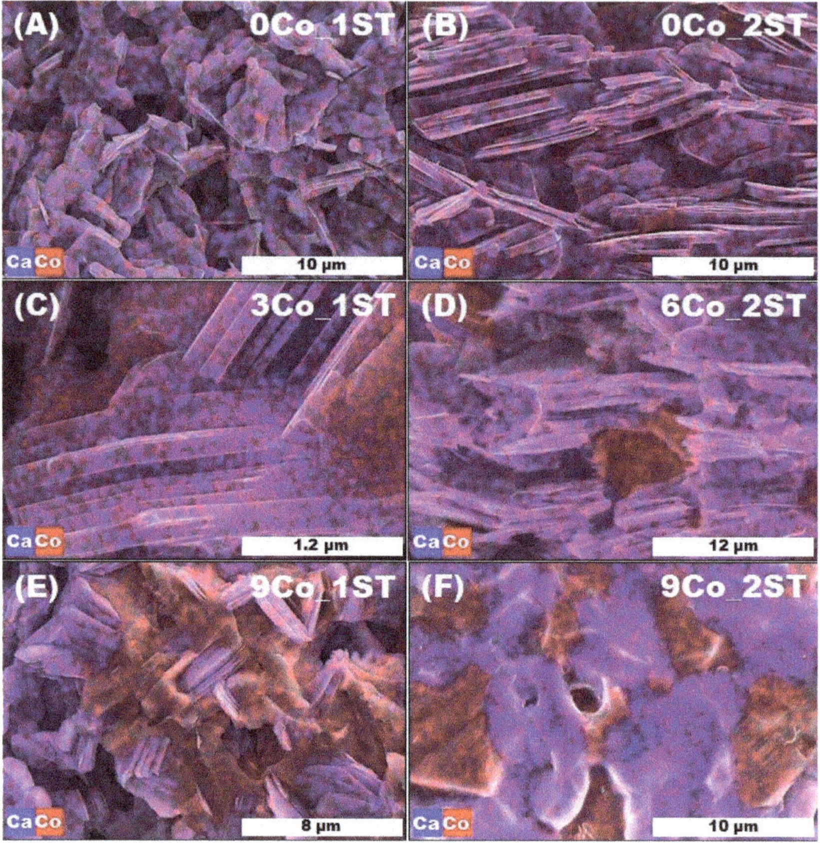

**Figure 4.** Representative EDS maps of fractured 1ST and 2ST sintered samples: (**A**) 0Co_1ST, (**B**) 0Co_2ST, (**C**) 3Co_1ST, (**D**) 6Co_2ST, (**E**) 9Co_1ST and (**F**) 9Co_2ST.

The effect of improved grain connection vanishes in the 2ST sintered samples, where high density is inherent due to processing at a higher temperature. In this case, the excess of cobalt leads to rather complex morphologies and microstructural features, determined by an interplay between distinct formation kinetics of $Ca_3Co_4O_9$, $Ca_3Co_2O_6$ and $Co_3O_4$, affecting their spatial distribution. In fact, the grain morphology of these phases becomes essentially similar for 9Co_2ST (Figure 4F), in contrast to 9Co_1ST (Figure 4E), where elongated $Ca_3Co_4O_9$ grains are preserved and interconnected by $Co_3O_4$ inclusions.

The results presented up to this point are in agreement with the following high-temperature electrical performances' results.

### 3.3. Electrical Performance

The evolution of the electrical conductivity ($\sigma$) with temperature for the 1ST sintered samples is shown in Figure 5A. All $\sigma$ values increased almost linearly with temperature, showing a typical semiconducting behavior ($d\sigma/dT \geq 0$), also found elsewhere in literature, for similar cases [45,50,54,60,61]. The high-temperature electrical conduction mechanism characteristic for $Ca_3Co_4O_9$ is thermally activated hole hoping [62]. The electrical conductivity of $Ca_3Co_4O_9$ is highly anisotropic [13] and is mainly governed by the holes from the *ab* plane [33], i.e., by the conductive hexagonal [$CoO_2$] layers.

The formation of elongated grains (with mostly random orientation) during conventional solid-state processing results in high porosity and poor grain interconnectivity.

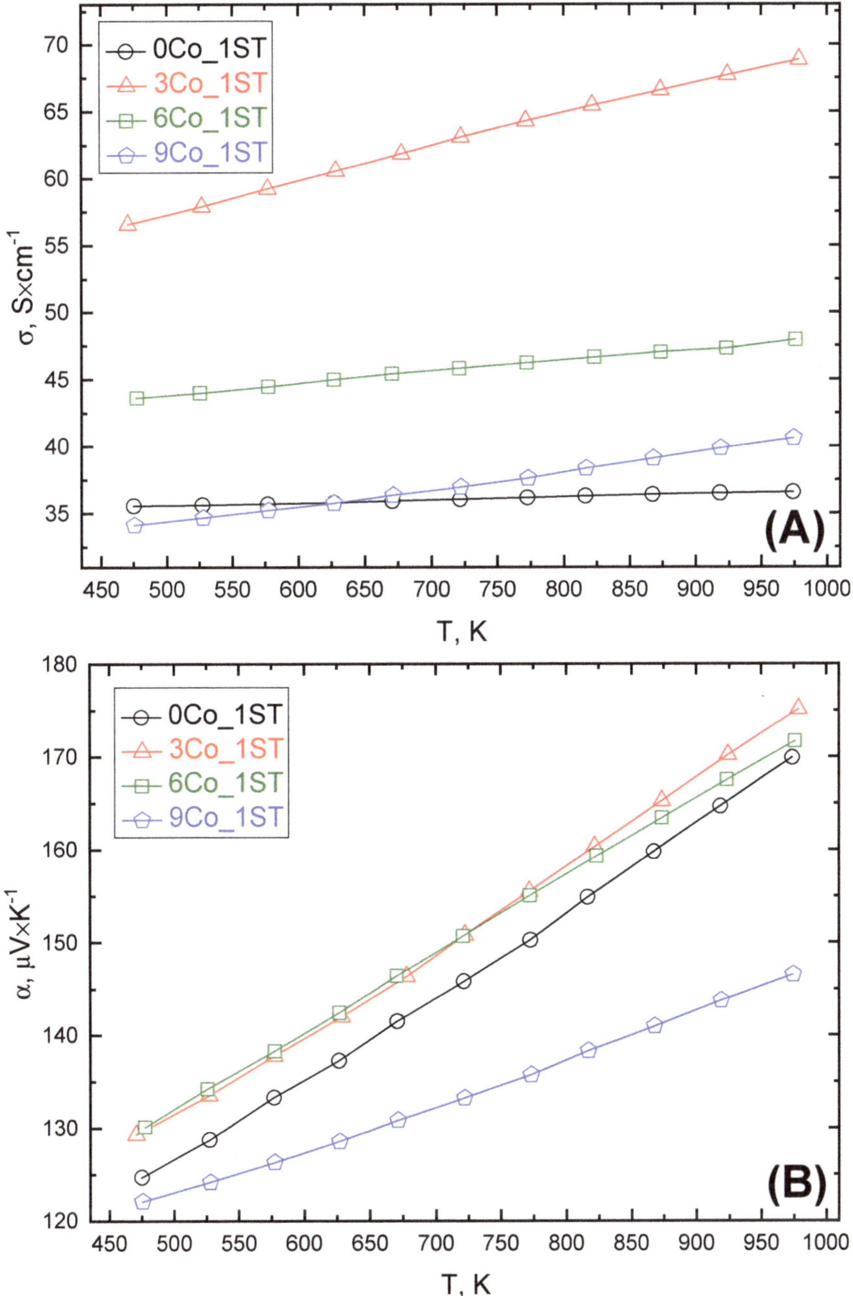

**Figure 5.** *Cont.*

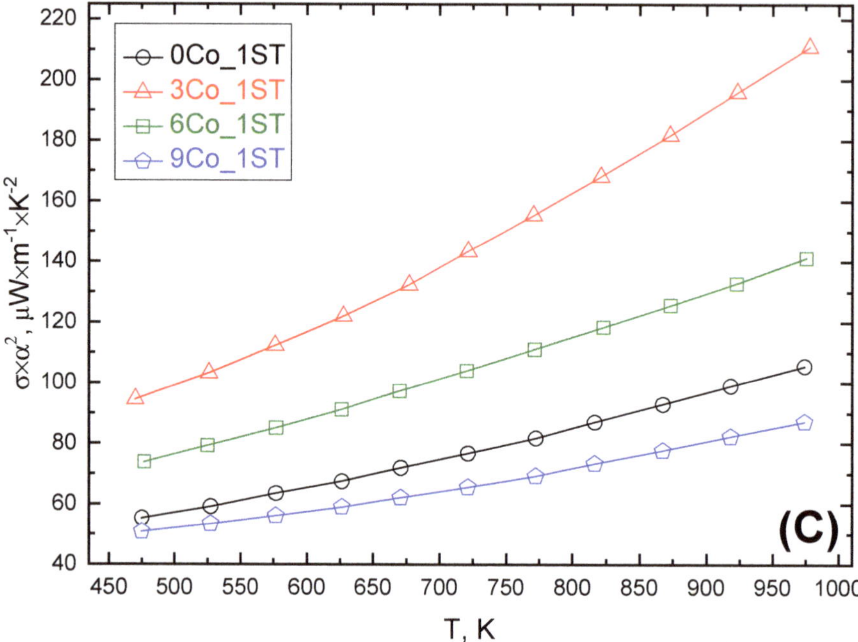

**Figure 5.** Electrical conductivity (**A**), Seebeck coefficient (**B**) and power factor (**C**) for 1ST sintered samples.

The obtained results unambiguously suggest that the proposed redox tailoring approach indeed results in a significant improvement of the charge carrier transport in the 1ST sintered materials. The electrical conductivity for the sample with the lowest amount of added cobalt, 3Co_1ST, is 1.6–1.9 times higher than for the pure matrix reference. It should also be noted that the highest $\sigma$ value of 68 Scm$^{-1}$, observed for 3Co_1ST at 975 K, is higher than some of the best-reported values in the literature [54,63]. This enhancement is believed to result from the improved grains interconnectivity, provided by the oxidation of added cobalt particles and filling of the pores with cobalt oxide. An oxidation-promoted filling by $Co_3O_4$ expansion in the pores is essential in this case, while direct sintering of $Ca_3Co_4O_9$ and $Co_3O_4$ mixture produces only a marginal improvement of the electrical conductivity in SPS processed samples, as found by F. Delorme [45]. It should be noticed that the electrical conductivity of 3Co_1ST with 61% of relative density (Table 1) is only slightly (~10% to 15%) lower than for 98% dense $Ca_3Co_4O_9$ prepared by SPS [45]. Even though $Co_3O_4$ intrinsically possesses notably lower total conductivity than $Ca_3Co_4O_9$ [45,64] and, thus, cannot provide a decisive enhancement of the charge carrier concentration, it contributes, however, by improving the charge carrier mobility, by connecting neighboring $Ca_3Co_4O_9$ grains, following the following basic equation [12]:

$$\sigma = ne\mu \tag{6}$$

where $n$, $e$ and $\mu$ are the charge carrier concentration, the electron charge and the carrier mobility, respectively. This hypothesis is also confirmed by comparable values of the Seebeck coefficient for 0Co_1ST, 3Co_1ST and 6Co_1ST; for the latter two composites, the thermopower is even slightly higher than for the pure matrix reference. This difference also agrees with the fact that the Seebeck coefficient of $Co_3O_4$ is noticeably higher compared to $Ca_3Co_4O_9$ [65], but appears counterintuitive when considering the significant drop in $\alpha$ observed for the 9Co_1ST sample (Figure 5B). A similar decrease was actually observed by F. Delorme [45] and attributed to the presence of the compressive strain originating from the mismatch of the thermal expansion coefficients of $Ca_3Co_4O_9$ and $Co_3O_4$, which is more likely to contribute at relatively higher $Co_3O_4$ contents, as those assessed in Reference [45]. The highest $\alpha$ values were measured for both 3 and 6Co_1ST, reaching the maximum value of 175 $\mu VK^{-1}$ at 975 K for 3Co_1ST, which is comparable to some of the best values reported in the literature [45,46].

The resulting PF values for the 1ST case are shown in Figure 5C. All samples demonstrate similar behavior: the PF values increase proportionally with temperature, in the whole measured temperature range. The PF values systematically decreased from 3Co_1ST to 9Co_1ST, following the corresponding trends observed for electrical conductivity and Seebeck coefficient. The highest PF value of 210 $\mu Wm^{-1} \cdot K^{-2}$, belonging to 3Co_1ST at 975 K, is among the best values reported in the literature, for highly textured, high-density samples [45,46].

Essentially opposite effects of the cobalt addition on the electrical transport properties can be observed for the denser 2ST sintered samples (Figure 6). In this case, all $\sigma$ values also increased linearly with temperature, in the whole measured temperature range, with the highest conductivity values measured for the reference 0Co_2ST sample. The lower $\sigma$ values measured for the cobalt-added samples can most probably be explained by the presence of the additional secondary phase, $Ca_3Co_2O_6$, besides $Co_3O_4$, known to possess $\sigma$ values of at least one order of magnitude lower than for $Ca_3Co_4O_9$ [66,67]. The lowest $\sigma$ values (between 13 and 28 $Scm^{-1}$) measured for 6 and 9Co_2ST are in agreement with the largest 23 and 40 wt.% fraction of $Ca_3Co_2O_6$, respectively, found in these samples. The reference sample 0Co_2ST has the largest $\sigma$ values (between 70 and 80 $Scm^{-1}$) measured in this work, due to its higher density (80% of $\rho_{th}$); this conductivity range is, again, comparable to some of the best-reported values from the literature [54,63]. In 3Co_2ST and 6Co_2ST, the cobalt addition also surprisingly resulted in a general decrease of the Seebeck coefficient (Figure 6B), which might be a result of various mechanical strains imposed by the complex phase composition. The combination of these factors resulted in a general decrease of the power factor measured for all cobalt-added 2ST sintered samples, as shown in Figure 6C.

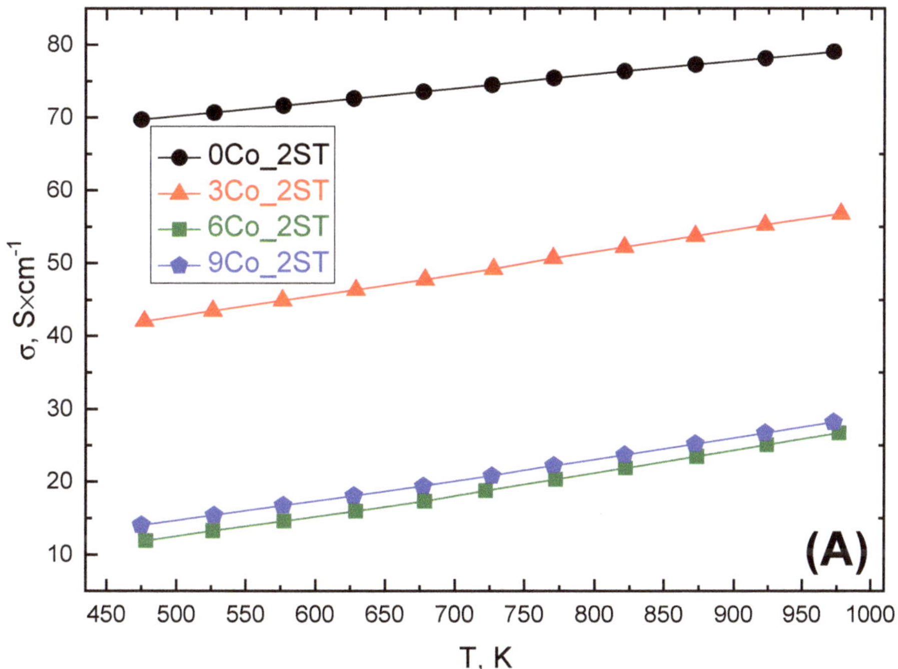


**Figure 6.** *Cont.*

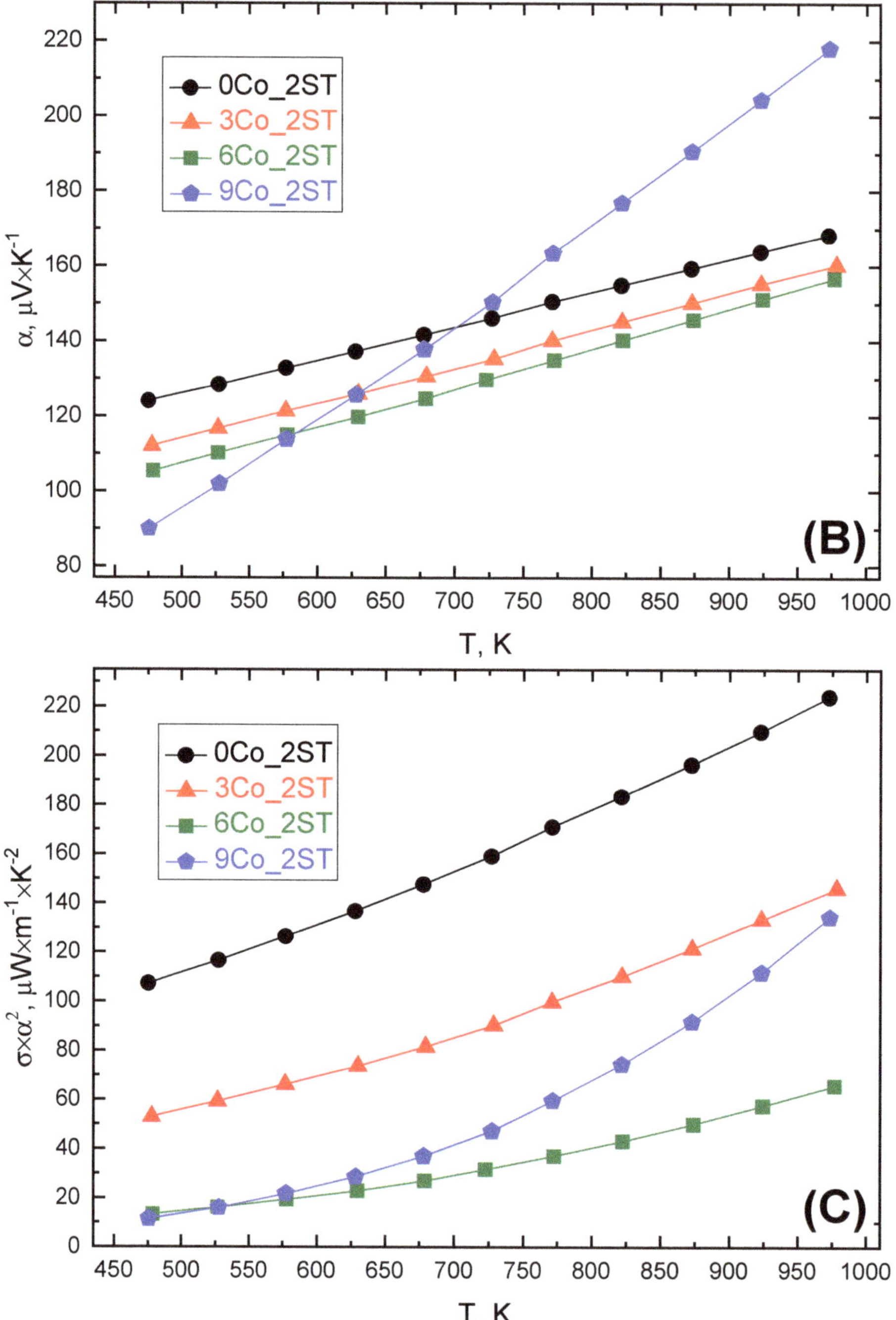

**Figure 6.** Electrical conductivity (**A**), Seebeck coefficient (**B**) and power factor (**C**) for 2ST sintered samples.

Finally, Figure 7 illustrates the compositional dependence of electrical performance at a fixed temperature.

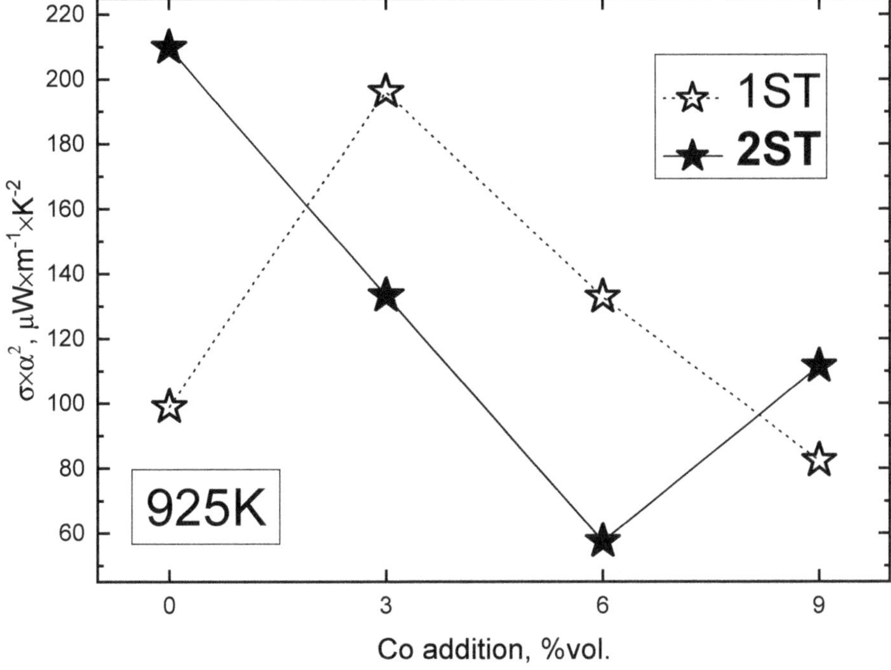


**Figure 7.** Compositional dependence of the power factor at 925 K, for the 1ST and 2ST sintered samples.

The highest PF values (around 200 $\mu Wm^{-1} \cdot K^{-2}$) from this work were achieved for 0Co_2ST and 3Co_1ST, due to the high density and grains interconnectivity improvement effect provided by the oxidation of cobalt, respectively. Higher cobalt additions promoted the formation of a larger amount of resistive secondary phases, which lead to lower PF values in the corresponding samples. Since a significant improvement of the electrical performance was found for a relatively low Co concentration, additional studies in the range of 1–5% vol. of cobalt addition might be necessary to further evaluate the potential of the proposed approach. The kinetics of pore-filling effects can be also adjusted by the size of the cobalt particles and the heating/dwell conditions. Thus, the scope of the present work includes the first demonstration of the redox-tailoring effects provided by the metal powder addition on the electrical counterpart of the TE performance of $Ca_3Co_4O_9$-based materials. Further assessment of the impact of the redox-induced interface on the thermal transport properties in such composite materials is also essential and needs to be addressed.

## 4. Conclusions

This work demonstrates a new processing route for tailoring the high-temperature electrical performance of $Ca_3Co_4O_9$-based materials, based on redox-promoted pore-filling effect on oxidation of metallic cobalt particles added to the $Ca_3Co_4O_9$ matrix. The efficacy of the proposed approach was shown to be strongly dependent on the processing conditions, resulting in a significant enhancement of the thermoelectric performance for relatively porous materials, where the effect of pore filling was more pronounced. This was provided by the presence of $Co_3O_4$ phase, which promoted an improvement in densification and better interconnectivity between $Ca_3Co_4O_9$ grains, simultaneously resulting in enhanced electrical transport. An alternative two-stage sintering route led to much denser $Ca_3Co_4O_9/Ca_3Co_2O_6/Co_3O_4$ ceramic samples, possessing, however, lower electrical performances due complex phase composition. The highest PF value of 210 $\mu Wm^{-1} \cdot K^{-2}$ at 975 K was observed for the $Ca_3Co_4O_9$-based ceramics containing 3% vol. of metallic cobalt addition, which is among the best-reported values in the literature for textured, high-density p-type layered cobaltites. The results highlight the importance of a proper grain boundaries design in $Ca_3Co_4O_9$-based thermoelectrics, where inherently poor densification during normal sintering can be compensated by introducing a suitable grain-connecting phase.

**Author Contributions:** Conceptualization, Gabriel Constantinescu, J.R.F. and A.V.K.; data curation, G.C. and A.R.S.; formal analysis, G.C., A.R.S. and A.V.K.; funding acquisition, J.R.F. and A.V.K.; investigation, G.C., A.R.S., S.R., D.L., S.S. and P.A.; methodology, G.C., A.R.S., S.R., D.L., S.S. and P.A.; project administration, A.V.K.; resources, J.R.F. and A.V.K.; supervision, J.R.F. and A.V.K.; validation, G.C. and A.R.S.; visualization, G.C. and A.V.K.; writing—original draft, G.C. and A.V.K.; writing—review and editing, G.C., A.R.S., S.R., D.L., S.S, P.A., J.R.F. and A.V.K. All authors have read and agreed to the published version of the manuscript.

**Funding:** This research was funded by the projects REMOTE (POCI-01-0145-FEDER-031875), CICECO–Aveiro Institute of Materials (UIDB/50011/2020 and UIDP/50011/2020) and SusPhotoSolutions (Centro-01-0145-FEDER-000005), financed by COMPETE 2020 Program and National Funds through the FCT/MEC and, when applicable, co-financed by FEDER under the PT2020 Partnership Agreement.

**Acknowledgments:** Shahed Rasekh acknowledges the support of the FCT–CEECIND/02608/2017 grant.

**Conflicts of Interest:** The authors declare no conflict of interest.

## References

1. Rowe, D.M. Thermoelectric waste heat recovery as a renewable energy source. *Int. J. Innov. Energy Syst.* **2006**, *1*, 13–23.
2. Petsagkourakis, I.; Tybrandt, K.; Crispin, X.; Ohkubo, I.; Satoh, N.; Mori, T. Thermoelectric materials and applications for energy harvesting power generation. *Sci. Technol. Adv. Mater.* **2018**, *19*, 836–862. [CrossRef] [PubMed]
3. Bell, L.E. Cooling, Heating, Generating Power, and Recovering Waste Heat with Thermoelectric Systems. *Science* **2008**, *321*, 1457–1461. [CrossRef] [PubMed]
4. Rowe, D.M. *Modules, Systems, and Applications in Thermoelectrics*, 1st ed.; CRC Press: Boca Raton, FL, USA, 2012; pp. 1–581. ISBN 978-0-429-08825-4.
5. Champier, D. Thermoelectric generators: A review of applications. *Energy Convers. Manag.* **2017**, *140*, 167–181. [CrossRef]
6. Vining, C.B. An inconvenient truth about thermoelectrics. *Nat. Mater.* **2009**, *8*, 83–85. [CrossRef]
7. Urban, J.J.; Menon, A.K.; Tian, Z.; Jain, A.; Hippalgaonkar, K. New horizons in thermoelectric materials: Correlated electrons, organic transport, machine learning, and more. *J. Appl. Phys.* **2019**, *125*, 180902. [CrossRef]
8. Alam, H.; Ramakrishna, S. A review on the enhancement of figure of merit from bulk to nano thermoelectric materials. *Nano Energy* **2013**, *2*, 190–212. [CrossRef]
9. Pei, Y.; Shi, X.; LaLonde, A.; Wang, H.; Chen, L.; Snyder, G.J. Convergence of electronic bands for high performance bulk thermoelectrics. *Nature* **2011**, *473*, 66–69. [CrossRef]
10. Pei, Y.; Wang, H.; Snyder, G.J. Band Engineering of Thermoelectric Materials. *Adv. Mater.* **2012**, *24*, 6125–6135. [CrossRef]
11. Hanus, R.; Agne, M.T.; Rettie, A.J.E.; Chen, Z.; Tan, G.; Chung, D.Y.; Kanatzidis, M.G.; Pei, Y.; Voorhees, P.W.; Snyder, G.J. Lattice Softening Significantly Reduces Thermal Conductivity and Leads to High Thermoelectric Efficiency. *Adv. Mater.* **2019**, *31*, 1900108. [CrossRef]
12. Snyder, G.J.; Toberer, E.S. Complex thermoelectric materials. *Nat. Mater.* **2008**, *7*, 105–114. [CrossRef] [PubMed]
13. Wolf, M.; Hinterding, R.; Feldhoff, A. High Power Factor vs. High zT—A Review of Thermoelectric Materials for High-Temperature Application. *Entropy* **2019**, *21*, 1058. [CrossRef]
14. LeBlanc, S. Thermoelectric generators: Linking material properties and systems engineering for waste heat recovery applications. *Sustain. Mater. Technol.* **2014**, *1–2*, 26–35. [CrossRef]
15. Terasaki, I.; Sasago, Y.; Uchinokura, K. Large thermoelectric power in $NaCo_2O_4$ single crystals. *Phys. Rev. B* **1997**, *56*, R12685–R12687. [CrossRef]
16. Ji, L. 3-Metal oxide-based thermoelectric materials. In *Metal Oxides in Energy Technologies. Metal Oxides*; Elsevier: Amsterdam, The Netherlands, 2018; pp. 49–72. ISBN 9780128111673.

17. Matsubara, I.; Funahashi, R.; Takeuchi, T.; Sodeoka, S.; Shimizu, T.; Ueno, K. Fabrication of an all-oxide thermoelectric power generator. *Appl. Phys. Lett.* **2001**, *78*, 3627–3629. [CrossRef]

18. Fergus, J.W. Oxide materials for high temperature thermoelectric energy conversion. *J. Eur. Ceram. Soc.* **2012**, *32*, 525–540. [CrossRef]

19. Saucke, G.; Populoh, S.; Thiel, P.; Xie, W.; Funahashi, R.; Weidenkaff, A. Compatibility approach for the improvement of oxide thermoelectric converters for industrial heat recovery applications. *J. Appl. Phys.* **2015**, *118*, 035106. [CrossRef]

20. Merkulov, O.V.; Politov, B.V.; Chesnokov, K.Y.; Markov, A.A.; Leonidov, I.A.; Patrakeev, M.V. Fabrication and Testing of a Tubular Thermoelectric Module Based on Oxide Elements. *J. Electron. Mater.* **2018**, *47*, 2808–2816. [CrossRef]

21. Bittner, M.; Kanas, N.; Hinterding, R.; Steinbach, F.; Räthel, J.; Schrade, M.; Wiik, K.; Einarsrud, M.-A.; Feldhoff, A. A comprehensive study on improved power materials for high-temperature thermoelectric generators. *J. Power Sour.* **2019**, *410–411*, 143–151. [CrossRef]

22. Kovalevsky, A.V.; Aguirre, M.H.; Populoh, S.; Patrício, S.G.; Ferreira, N.M.; Mikhalev, S.M.; Fagg, D.P.; Weidenkaff, A.; Frade, J.R. Designing strontium titanate-based thermoelectrics: Insight into defect chemistry mechanisms. *J. Mater. Chem. A* **2017**, *5*, 3909–3922. [CrossRef]

23. Xiao, X.; Widenmeyer, M.; Mueller, K.; Scavini, M.; Checchia, S.; Castellano, C.; Ma, D.; Yoon, S.; Xie, W.; Starke, U.; et al. A squeeze on the perovskite structure improves the thermoelectric performance of Europium Calcium Titanates. *Mater. Today Phys.* **2018**, *7*, 96–105. [CrossRef]

24. Thébaud, S.; Adessi, C.; Bouzerar, G. Investigating the high-temperature thermoelectric properties of n-type rutile $TiO_2$. *Phys. Rev. B* **2019**, *100*, 195202. [CrossRef]

25. Moos, R.; Hardtl, K.H. Defect Chemistry of Donor-Doped and Undoped Strontium Titanate Ceramics between 1000 °C and 1400 °C. *J. Am. Ceram. Soc.* **2005**, *80*, 2549–2562. [CrossRef]

26. Maignan, A.; Hébert, S.; Pi, L.; Pelloquin, D.; Martin, C.; Michel, C.; Hervieu, M.; Raveau, B. Perovskite manganites and layered cobaltites: Potential materials for thermoelectric applications. *Cryst. Eng.* **2002**, *5*, 365–382. [CrossRef]

27. Ferreira, N.M.; Neves, N.R.; Ferro, M.C.; Torres, M.A.; Madre, M.A.; Costa, F.M.; Sotelo, A.; Kovalevsky, A.V. Growth rate effects on the thermoelectric performance of $CaMnO_3$-based ceramics. *J. Eur. Ceram. Soc.* **2019**, *39*, 4184–4188. [CrossRef]

28. Merkulov, O.V.; Patrakeev, M.V.; Leonidov, I.A. Electrical Transport Properties of $Ca_{1-x}Bi_xMnO_{3-\delta}$. *Inorg. Mater.* **2019**, *55*, 1014–1019. [CrossRef]

29. Sotelo, A.; Depriester, M.; Torres, M.A.; Sahraoui, A.H.; Madre, M.A.; Diez, J.C. Effect of simultaneous K, and Yb substitution for Ca on the microstructural and thermoelectric characteristics of $CaMnO_3$ ceramics. *Ceram. Int.* **2018**, *44*, 12697–12701. [CrossRef]

30. Zakharchuk, K.V.; Widenmeyer, M.; Alikin, D.O.; Xie, W.; Populoh, S.; Mikhalev, S.M.; Tselev, A.; Frade, J.R.; Weidenkaff, A.; Kovalevsky, A.V. A self-forming nanocomposite concept for ZnO-based thermoelectrics. *J. Mater. Chem. A* **2018**, *6*, 13386–13396. [CrossRef]

31. Tsubota, T.; Ohtaki, M.; Eguchi, K.; Arai, H. Thermoelectric properties of Al-doped ZnO as a promising oxide material for high-temperature thermoelectric conversion. *J. Mater. Chem.* **1997**, *7*, 85–90. [CrossRef]

32. Ohtaki, M.; Tsubota, T.; Eguchi, K.; Arai, H. High-temperature thermoelectric properties of $(Zn_{1-x}Al_x)O$. *J. Appl. Phys.* **1996**, *79*, 1816–1818. [CrossRef]

33. Masset, A.C.; Michel, C.; Maignan, A.; Hervieu, M.; Toulemonde, O.; Studer, F.; Raveau, B.; Hejtmanek, J. Misfit-layered cobaltite with an anisotropic giant magnetoresistance: $Ca_3Co_4O_9$. *Phys. Rev. B* **2000**, *62*, 166–175. [CrossRef]

34. Miyazaki, Y.; Onoda, M.; Oku, T.; Kikuchi, M.; Ishii, Y.; Ono, Y.; Morii, Y.; Kajitani, T. Modulated Structure of the Thermoelectric Compound $[Ca_2CoO_3]_{0.62}CoO_2$. *J. Phys. Soc. Jpn.* **2002**, *71*, 491–497. [CrossRef]

35. Takada, K.; Sakurai, H.; Takayama-Muromachi, E.; Izumi, F.; Dilanian, R.A.; Sasaki, T. Superconductivity in two-dimensional $CoO_2$ layers. *Nature* **2003**, *422*, 53–55. [CrossRef] [PubMed]

36. Maignan, A.; Pelloquin, D.; Hebert, S.; Klein, Y.; Hervieu, M. Potencia termoeléctrica de cerámicas basadas en cobaltitas: Optimización mediante sustitución química (Thermoelectric Power in Misfit Cobaltites Ceramics: Optimization by Chemical Substitutions). *Bol. Soc. Esp. Ceram. Vidr.* **2006**, *45*, 122–125. [CrossRef]

37. Constantinescu, G.; Torres, M.A.; Rasekh, S.H.; Bosque, P.; Madre, M.A.; Diez, J.C.; Sotelo, A. Thermoelectric properties in $Ca_3Co_{4-x}Mn_xO_y$ ceramics. *Adv. Appl. Ceram.* **2015**, *114*, 303–308. [CrossRef]

38. Butt, S.; Xu, W.; He, W.Q.; Tan, Q.; Ren, G.K.; Lin, Y.; Nan, C.-W. Enhancement of thermoelectric performance in Cd-doped $Ca_3Co_4O_9$ via spin entropy, defect chemistry and phonon scattering. *J. Mater. Chem. A* **2014**, *2*, 19479–19487. [CrossRef]

39. Boyle, C.; Liang, L.; Chen, Y.; Prucz, J.; Cakmak, E.; Watkins, T.R.; Lara-Curzio, E.; Song, X. Competing dopants grain boundary segregation and resultant seebeck coefficient and power factor enhancement of thermoelectric calcium cobaltite ceramics. *Ceram. Int.* **2017**, *43*, 11523–11528. [CrossRef]

40. Miyazawa, K.; Amaral, F.; Kovalevsky, A.V.; Graça, M.P.F. Hybrid microwave processing of $Ca_3Co_4O_9$ thermoelectrics. *Ceram. Int.* **2016**, *42*, 9482–9487. [CrossRef]

41. Torres, M.A.; Costa, F.M.; Flahaut, D.; Touati, K.; Rasekh, S.H.; Ferreira, N.M.; Allouche, J.; Depriester, M.; Madre, M.A.; Kovalevsky, A.V.; et al. Significant enhancement of the thermoelectric performance in $Ca_3Co_4O_9$. Thermoelectric materials through combined strontium substitution and hot-pressing process. *J. Eur. Ceram. Soc.* **2019**, *39*, 1186–1192. [CrossRef]

42. Bresch, S.; Mieller, B.; Schoenauer-Kamin, D.; Moos, R.; Giovanelli, F.; Rabe, T. Influence of pressure assisted sintering and reaction sintering on microstructure and thermoelectric properties of bi-doped and undoped calcium cobaltite. *J. Appl. Phys.* **2019**, *126*, 075102. [CrossRef]

43. Bergman, D.J.; Fel, L.G. Enhancement of thermoelectric power factor in composite thermoelectrics. *J. Appl. Phys.* **1999**, *85*, 8205–8216. [CrossRef]

44. Faleev, S.V.; Léonard, F. Theory of enhancement of thermoelectric properties of materials with nanoinclusions. *Phys. Rev. B* **2008**, *77*, 214304. [CrossRef]

45. Delorme, F.; Diaz-Chao, P.; Guilmeau, E.; Giovannelli, F. Thermoelectric properties of $Ca_3Co_4O_9$–$Co_3O_4$ composites. *Ceram. Int.* **2015**, *41*, 10038–10043. [CrossRef]

46. Noudem, J.G.; Kenfaui, D.; Chateigner, D.; Gomina, M. Granular and Lamellar Thermoelectric Oxides Consolidated by Spark Plasma Sintering. *J. Electron. Mater.* **2011**, *40*, 1100–1106. [CrossRef]

47. Kenfaui, D.; Bonnefont, G.; Chateigner, D.; Fantozzi, G.; Gomina, M.; Noudem, J.G. $Ca_3Co_4O_9$ ceramics consolidated by SPS process: Optimisation of mechanical and thermoelectric properties. *Mater. Res. Bull.* **2010**, *45*, 1240–1249. [CrossRef]

48. Madre, M.A.; Costa, F.M.; Ferreira, N.M.; Sotelo, A.; Torres, M.A.; Constantinescu, G.; Rasekh, S.H.; Diez, J.C. Preparation of high-performance $Ca_3Co_4O_9$ thermoelectric ceramics produced by a new two-step method. *J. Eur. Ceram. Soc.* **2013**, *33*, 1747–1754. [CrossRef]

49. Sotelo, A.; Constantinescu, G.; Rasekh, Sh.; Torres, M.A.; Diez, J.C.; Madre, M.A. Improvement of thermoelectric properties of $Ca_3Co_4O_9$ using soft chemistry synthetic methods. *J. Eur. Ceram. Soc.* **2012**, *32*, 2415–2422. [CrossRef]

50. Bittner, M.; Helmich, L.; Nietschke, F.; Geppert, B.; Oeckler, O.; Feldhoff, A. Porous $Ca_3Co_4O_9$ with enhanced thermoelectric properties derived from Sol–Gel synthesis. *J. Eur. Ceram. Soc.* **2017**, *37*, 3909–3915. [CrossRef]

51. Królicka, A.K.; Piersa, M.; Mirowska, A.; Michalska, M. Effect of sol-gel and solid-state synthesis techniques on structural, morphological and thermoelectric performance of $Ca_3Co_4O_9$. *Ceram. Int.* **2018**, *44*, 13736–13743. [CrossRef]

52. Woermann, E.; Muan, A. Phase equilibria in the system CaO-cobalt oxide in air. *J. Inorg. Nucl. Chem.* **1970**, *32*, 1455–1459. [CrossRef]

53. Sedmidubský, D.; Jakeš, V.; Jankovský, O.; Leitner, J.; Sofer, Z.; Hejtmánek, J. Phase equilibria in Ca–Co–O system. *J. Solid State Chem.* **2012**, *194*, 199–205. [CrossRef]

54. Schulz, T.; Töpfer, J. Thermoelectric properties of $Ca_3Co_4O_9$ ceramics prepared by an alternative pressure-less sintering/annealing method. *J. Alloy Compd.* **2016**, *659*, 122–126. [CrossRef]

55. Kovalevsky, A.V.; Yaremchenko, A.A.; Populoh, S.; Weidenkaff, A.; Frade, J.R. Enhancement of thermoelectric performance in strontium titanate by praseodymium substitution. *J. Appl. Phys.* **2013**, *113*, 053704. [CrossRef]

56. Patrakeev, M.V.; Mitberg, E.B.; Lakhtin, A.A.; Leonidov, I.A.; Kozhevnikov, V.L.; Kharton, V.V.; Avdeev, M.; Marques, F.M.B. Oxygen Nonstoichiometry, Conductivity, and Seebeck Coefficient of $La_{0.3}Sr_{0.7}Fe_{1-x}Ga_xO_{2.65+\delta}$ Perovskites. *J. Solid State Chem.* **2002**, *167*, 203–213. [CrossRef]

57. Gulbransen, E.A.; Andrew, K.F. The Kinetics of the Oxidation of Cobalt. *J. Electrochem. Soc.* **1951**, *98*, 241. [CrossRef]

58. Büttner, G.; Populoh, S.; Xie, W.; Trottmann, M.; Hertrampf, J.; Döbeli, M.; Karvonen, L.; Yoon, S.; Thiel, P.; Niewa, R.; et al. Thermoelectric properties of $[Ca_2CoO_{3-\delta}][CoO_2]_{1,62}$ as a function of Co/Ca defects and $Co_3O_4$ inclusions. *J. Appl. Phys.* **2017**, *121*, 215101. [CrossRef]

59. Liou, Y.C.; Tsai, W.C.; Lin, W.Y.; Lee, U.R. Synthesis of $Ca_3Co_4O_9$ and $CuAlO_2$ ceramics of the thermoelectric application using a reaction sintering process. *J. Aust. Ceram. Soc.* **2008**, *44*, 17–22.

60. Kahraman, F.; Madre, M.A.; Rasekh, S.; Salvador, C.; Bosque, P.; Torres, M.A.; Diez, J.C.; Sotelo, A. Enhancement of mechanical and thermoelectric properties of $Ca_3Co_4O_9$ by Ag addition. *J. Eur. Ceram. Soc.* **2015**, *35*, 3835–3841. [CrossRef]

61. Presečnik, M.; de Boor, J.; Bernik, S. Synthesis of single-phase $Ca_3Co_4O_9$ ceramics and their processing for a microstructure-enhanced thermoelectric performance. *Ceram. Int.* **2016**, *42*, 7315–7327. [CrossRef]

62. Lin, Y.-H.; Lan, J.; Shen, Z.; Liu, Y.; Nan, C.-W.; Li, J.-F. High-temperature electrical transport behaviors in textured $Ca_3Co_4O_9$-based polycrystalline ceramics. *Appl. Phys. Lett.* **2009**, *94*, 072107. [CrossRef]

63. Zhou, Y.; Matsubara, I.; Horii, S.; Takeuchi, T.; Funahashi, R.; Shikano, M.; Shimoyama, J.; Kishio, K.; Shin, W.; Izu, N.; et al. Thermoelectric properties of highly grain-aligned and densified Co-based oxide ceramics. *J. Appl. Phys.* **2003**, *93*, 2653–2658. [CrossRef]

64. Bordeneuve, H.; Guillemet-Fritsch, S.; Rousset, A.; Schuurman, S.; Poulain, V. Structure and electrical properties of single-phase cobalt manganese oxide spinels $Mn_{3-x}Co_xO4$ sintered classically and by spark plasma sintering (SPS). *J. Solid State Chem.* **2009**, *182*, 396–401. [CrossRef]

65. Broemme, A.D.D. Correlation between thermal expansion and Seebeck coefficient in polycrystalline $Co_3O_4$. *IEEE Trans. Elect. Insul.* **1991**, *26*, 49–52. [CrossRef]

66. Maignan, A.; Hébert, S.; Martin, C.; Flahaut, D. One dimensional compounds with large thermoelectric power: $Ca_3Co_2O_6$ and $Ca_3CoMO_6$ with M = $Ir^{4+}$ and $Rh^{4+}$. *Mater. Sci. Eng. B* **2003**, *104*, 121–125. [CrossRef]

67. Iwasaki, K.; Yamane, H.; Kubota, S.; Takahashi, J.; Shimada, M. Power factors of $Ca_3Co_2O_6$ and $Ca_3Co_2O_6$-based solid solutions. *J. Alloy Compd.* **2003**, *358*, 210–215. [CrossRef]

 *materials* 

*Article*

# Metallization and Diffusion Bonding of CoSb$_3$-Based Thermoelectric Materials

Hangbin Feng [1], Lixia Zhang [1,*], Jialun Zhang [1], Wenqin Gou [1], Sujuan Zhong [2],
Guanxing Zhang [2], Huiyuan Geng [1,*] and Jicai Feng [1]

[1]   State Key Laboratory of Advanced Welding and Joining, Harbin Institute of Technology,
      Harbin 150001, China; 18245017507@163.com (H.F.); zhangjialunhit@163.com (J.Z.);
      19S009143@stu.hit.edu.cn (W.G.); fjc_hit@163.com (J.F.)
[2]   Zhengzhou Research Institute of Mechanical Engineering, Zhengzhou 450001, China; zsj_hit@163.com (S.Z.);
      zgx_hit1@163.com (G.Z.)
*    Correspondence: zhanglxia@hit.edu.cn (L.Z.); genghuiyuan@hit.edu.cn (H.G.)

Received: 9 January 2020; Accepted: 26 February 2020; Published: 3 March 2020

**Abstract:** CoSb$_3$-based skutterudite alloy is one of the most promising thermoelectric materials in the middle temperature range (room temperature—550 °C). However, the realization of an appropriate metallization layer directly on the sintered skutterudite pellet is indispensable for the real thermoelectric generation application. Here, we report an approach to prepare the metallization layer and the subsequent diffusion bonding method for the high-performance multi-filled $n$-type skutterudite alloys. Using the electroplating followed by low-temperature annealing approaches, we successfully fabricated a Co-Mo metallization layer on the surface of the skutterudite alloy. The coefficient of thermal expansion of the electroplated layer was optimized by changing its chemical composition, which can be controlled by the electroplating temperature, current and the concentration of the Mo ions in the solution. We then joined the metallized skutterudite leg to the Cu-Mo electrode using a diffusion bonding method performed at 600 °C and 1 MPa for 10 min. The Co-Mo/skutterudite interfaces exhibit extremely low specific contact resistivity of 1.41 $\mu\Omega$ cm$^2$. The metallization layer inhibited the elemental inter-diffusion to less than 11 $\mu$m after annealing at 550 °C for 60 h, indicating a good thermal stability. The current results pave the way for the large-scale fabrication of CoSb$_3$-based thermoelectric modules.

**Keywords:** thermoelectric materials; joining; skutterudite alloy; Co-Mo metallization layer

## 1. Introduction

Thermoelectric materials have attracted great attention for centuries because they can directly convert heat to electricity, and vice versa [1]. Nowadays, thermoelectric generation is also an important green energy technology for solving the energy crisis [2]. Among many thermoelectric materials, the skutterudite (SKU) compound is one of the most promising thermoelectric materials for the middle temperature range of thermoelectric generation. The skutterudite compound not only possesses high thermoelectric performance, but also good mechanical properties and excellent economic and environmental friendliness [3–8].

In the fabrication of a thermoelectric device, both $p$-type and $n$-type thermoelectric materials are required to be connected with the metal electrodes [9]. However, due to problems such as the large contact resistance, elemental diffusion and coefficient of thermal expansion (CTE) mismatch, how to connect the skutterudites to the electrodes reliably is still a major challenge [10]. Copper and nickel are the common electrode materials for the thermoelectric devices. However, when the electrode connects to the skutterudite compounds directly, they react violently. This reaction results in a great degradation of the thermoelectric performance of the skutterudite compounds [11,12]. Therefore, it is indispensable to prepare a diffusion barrier layer on the surface of the skutterudite compound. At present, most of the diffusion barrier layers of thermoelectric materials are fabricated by the sintering method [13–15]. The connection is achieved using the brazing method [16]. For example, Guo et al. used Fe-Co-Ni alloys as the diffusion barriers suitable for their *p*- and *n*-type skutterudite compounds [16]. Co, Fe, Ni and some minor elements were arc melted into buttons. Round disks were then sliced from the buttons and sintered on the top and bottom sides of the *n*- and *p*-type pellets to serve as diffusion barriers. After that, the metallized pellets were brazed to the Cu electrodes using the Ag-Cu-Zn solders [16–18]. Gu et al. used the Ti-Al mixed powders as the diffusion barrier. The Ti-Al mixed powders were put on the top of the $Yb_{0.3}Co_4Sb_{12}$ fine powders during the bonding process of the spark plasma sintering (SPS); the Ti powders reacted with Sb and Co to form a thin layer of intermetallic compound, which provided necessary bonding strength [19]. However, when the diffusion barrier layers were sintered with the thermoelectric powders, both the sintering temperature and pressure were limited by the physical properties of the thermoelectric materials; the barrier layer was often porous, resulting in the large contact resistivity of the thermoelectric leg. Furthermore, the preparation of the barrier layer by the sintering method often resulted in the difficulty of controlling the size of the thermoelectric leg accurately. The size of the sintered pellets was also limited by the thermal stresses caused by the CTE mismatch. Thus, these features were inconvenient for the large-scale device fabrication.

Electroplating metallic layers on thermoelectric materials followed by annealing is widely used in the thermoelectric community [20]. It has been reported that a cobalt barrier layer was prepared on the surface of a skutterudite leg by the electroplating method. However, since the CTE of cobalt is larger than that of the skutterudite alloy, the joint was easy to crack [21]. It was also reported that a Ti barrier layer on the surface of the skutterudite alloy was prepared by the magnetron sputtering approach [22]. The CTE of Ti was similar to that of the skutterudite alloy; but the diffusion ability of Ti in the skutterudite alloy was very weak, resulting in low bonding strength between them [23]. Zhao et al. prepared a Ti barrier layer by the SPS method and systematically studied the diffusion kinetics of the $Ti/CoSb_3$ interface at high temperatures. Using the interfacial shearing strength as the criteria, Zhao et al. also predicted a service life of over 20 years for the $Ti/CoSb_3$ joints at 500 °C [24,25].

It is clear that the realization of an appropriate diffusion barrier layer directly on the sintered skutterudite pellet is critical for the real thermoelectric generation application. Here, we first synthesized multifilled *n*-type skutterudite pellets. Then, the Co-Mo nanograined layers were fabricated on the surface of the pellets by the traditional electroplating and subsequent low-temperature annealing approaches. Both the CTE and the elemental diffusion behaviors were controlled by tuning the Mo content in the layer. Finally, a reliable joint between the skutterudite alloy and the Cu-Mo electrode was achieved using the diffusion bonding method.

## 2. Materials and Methods

### 2.1. Fabrication of the Skutterudite Alloys

Yb ingot (99.9%, Alfa Aesar), Al ingot (99.9%, Alfa Aesar), Ga ingot (99.99%, Alfa Aesar), In ingot (99.9%, Alfa Aesar), Co ingot (99.95%, Alfa Aesar), Ca granules (99.5%, Alfa Aesar), Fe granules (99.98%, Alfa Aesar) and Sb balls (99.999%, Alfa Aesar, Ward Hill, US) were weighted according to the chemical composition of $Yb_{0.3}Ca_{0.1}Al_{0.1}Ga_{0.1}In_{0.1}Co_4Sb_{12}$. The weighted elements were loaded into a carbon crucible and then were sealed in the quartz tubes under vacuum below $10^{-3}$ Pa. The quartz tubes were slowly heated to 1150 °C, held at this temperature for 5 h and then quenched in cold water. The quenched ingots were annealed at 700 °C for 150 h. The annealed samples were ground into fine powders in sizes of < 75 μm in a glovebox filled with high-purity Ar. Finally, the powders were sintered using SPS (Dr. Sinter Lab, Tsurugashima, Japan) at 700 °C for 15 min under a pressure of 60 MPa.

## 2.2. The Electroplating Process

The bulk skutterudite pellets were cut into cubes in the size of 5 mm × 5 mm × 5 mm using a diamond wire saw (STX-202A, Shenyang, China). The surfaces of the cubes were then polished. Before electroplating, the cubes were first placed in acetone for ultrasonic cleaning for 10 min to remove the oil stain on the surface of the material. During the plating of Co-Mo alloy on the surface of the skutterudite cubes, the temperature of the plating solution was set to 30–60 °C; the pH of the plating solution was controlled at 5.0–7.0 and the current was set to 0.07–0.16 A. It took about 30 min for the electroplating process. The plated skutterudite cubes were then sealed in the quartz tubes under vacuum below $10^{-3}$ Pa. Thereafter, the quartz tubes were placed in a box furnace for annealing at 450 °C for 2 h. Finally, the metallized skutterudite cubes were diffusion bonded to the Cu-Mo electrode. The applied pressure was about 1 MPa, and the joints were kept at 600 °C for 10 min to complete the diffusion bonding. To study the thermal stability of the metallization layer, we further annealed the joint at 550 °C in vacuum for 2 h, 5 h, 50 h, and 60 h, respectively.

## 2.3. Characterizations

Electrical transport properties of the obtained skutterudite alloys, including electrical conductivity and the Seebeck coefficient, were measured using the CTA-3 (Cryoall Co. Ltd, Beijing, China) apparatus. The maximum $ZT$ of the $n$-type skutterudite compound was about 1.35@773K. The details of the thermoelectric properties of the compound will be reported in our future manuscripts. The interfacial microstructure of the joints was investigated by a scanning electron microscope (Quanta 200FEG and Merlin Compact, Thermo Fisher Scientific, Waltham, US), and the chemical composition of the interfacial diffusion layers was analyzed using an energy dispersive spectrometer (EDS). The specific contact resistivity of the joints was measured on homemade four-probe equipment. The uncertainty of the homemade equipment is 5%.

## 3. Results and Discussion

### 3.1. Metallization of Skutterudite Alloys

Since the skutterudite-based thermoelectric generators usually work between 50 °C and 600 °C, it is necessary to prepare a barrier layer on the surface of the skutterudite. Otherwise, the Cu-based electrodes will react violently with the skutterudite leg, leading to the lower joint strength and thermoelectric performance of the generator. Generally speaking, the choice of the barrier layer needs to meet the following criteria: (i) The barrier layer can simultaneously form a finite reaction layer with the skutterudite compound and the electrode to form a high-strength joint. (ii) The barrier layer should possess a similar CTE to that of the skutterudite compound and the metal electrode. (iii) The interfacial electrical and thermal resistances caused by the introduction of the barrier layer should be as low as possible. According to the above criteria, the Co-Mo alloy was chosen as the barrier layer between the skutterudite compound and the Cu-Mo electrodes.

Figure 1a shows the cross-section of the electroplated layer. It is clear that the electroplated layer was just mechanically bonded to the surface of the skutterudite compound. Some parts of the electroplated barrier layer were separated from the skutterudite just after the electroplating process. The bonding strength between the barrier layer and the skutterudite alloy was far from the requirement of an ideal connection. To ensure the bonding strength, subsequent annealing under vacuum was carried out. As shown in Figure 1b, after annealing at 450 °C for 2 h, a metallurgical bonding between the barrier layer and the skutterudite compound was achieved. Figure 1c shows the high-magnification backscattered electron (BSE) cross-section image of the interface between the skutterudite alloy and the electroplated layer. A thin reaction layer in a thickness of 1 μm formed just on the top of the skutterudite compound. Some nanograined columnar crystals inside the barrier layer can also be observed clearly. The nanostructures in the electroplated layer introduced a vast amount of highly defective grain boundaries, which greatly reduced the diffusion energy barrier in the electroplated layer. Thus, a metallurgical bonding could be achieved at the low annealing temperature of 450 °C.

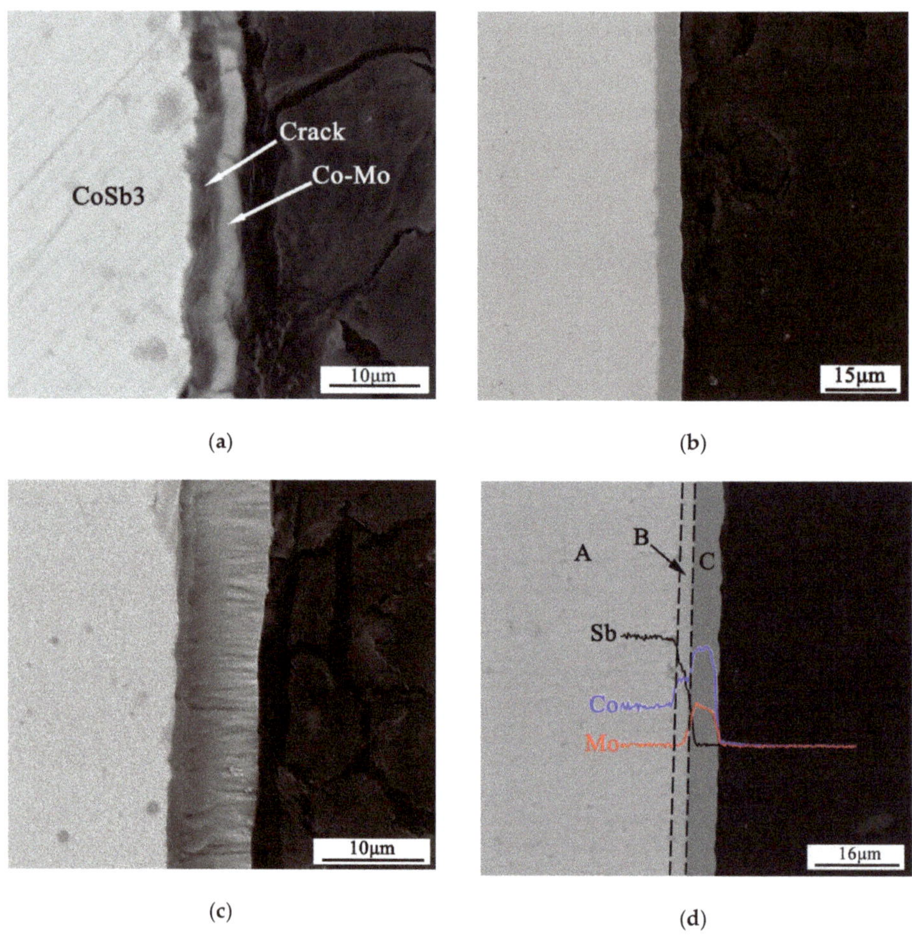


**Figure 1.** The backscattered electron (BSE) micrograph of Co-Mo/skutterudite (SKU) couple: (**a**) Before annealing, some parts of the electroplated barrier layer were separated from the skutterudite. (**b**) Annealed at 450 °C for 2 h, a metallurgical bonding between the barrier layer and the skutterudite compound was achieved. (**c**) The columnar crystals of the Co-Mo barrier layer. (**d**) The Co-Mo/SKU couple reacted at 500 °C for 5h. $CoSb_2$ of 3 μm was formed between the electroplated layer and skutterudite compound.

To gain insight into the reaction mechanism of the interfacial layer after annealing, we further increased the annealing temperature and time to obtain a thicker reaction layer. Figure 1d shows the cross-section microstructure and the EDS line-scanning results of the interface between the skutterudite compound and the electroplated layer after annealing at 500 °C for 5 h. According to the EDS results listed in Table 1, the Sb atoms diffused into the electroplated layer and $CoSb_2$ was formed between the electroplated layer and the skutterudite compound. The thickness of the reaction layer increased to about 3 μm.

Table 1. Analysis of interface components (at %).

|   | Co | Sb | Mo | Possible Phase |
|---|---|---|---|---|
| A | 26.10 | 73.87 | 0.03 | $CoSb_3$ |
| B | 33.01 | 66.93 | 0.06 | $CoSb_2$ |
| C | 74.90 | 0.25 | 24.85 | Co-Mo intermetallic phases |

### *3.2. Tuning the CTE of the Barrier Layer*

To reduce the thermal stresses between the thermoelectric materials and the electrode, the diffusion barrier should possess CTE close to those of the skutterudite compounds from room temperature to 600 °C. Guo et al. pointed out that the mismatch ratio of their CTEs should be limited to less than 10% [16]. The CTE of skutterudite is difficult to change, but the CTEs of Co-Mo alloys can be easily controlled by adjusting the chemical composition, as shown in Figure 2a. In fact, Song et al. demonstrated that the $Co_{0.6}Mo_{0.4}$ powder-mixed composite can serve well as a diffusion barrier for their *n*-type skutterudite compounds [26].

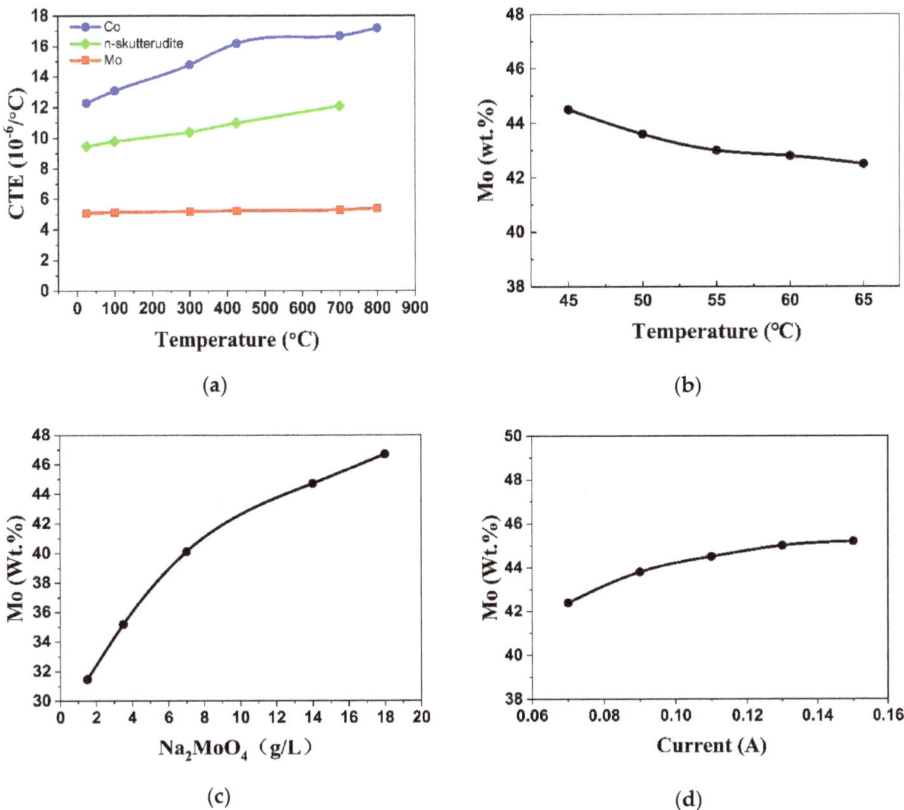

**Figure 2.** (**a**) The temperature-dependent coefficient of thermal expansion (CTE) of Co, skutterudite compound and Mo. (**b**) The effects of temperature on the Mo content of the electroplated layer. (**c**) The effects of the $Na_2MoO_4$ concentration on the Mo content of the electroplated layer. (**d**) The effects of the current on the Mo content of the electroplated layer.

Our results indicate that the Mo content in the Co-Mo barrier layer can be adjusted by changing the conditions of plating such as the temperature, current and composition of the plating solution. Figure 2b shows that when the current is 0.08 A and the content of $Na_2MoO_4$ in the plating solution is 8 g/L, the Mo concentration decreases slightly with the increasing temperature. On the contrary, when the current is 0.08A and the temperature is 65 °C, the content of Mo in the electroplated layer increases remarkably as the concentration of $Na_2MoO_4$ in the plating solution increases (Figure 2c). As for the current, the Mo content in the electroplated layer increases slightly with the increasing current when the temperature is about 65 °C and the content of $Na_2MoO_4$ in the plating solution is 8 g/L (as shown in Figure 2d).

The current can affect not only the composition of the electroplated layer, but also the surface morphology of the electroplated layer. As shown in Figure 3, the surface of the Co-Mo electroplated layer contains many nanosized particles. As the current increases, some cracks appear in the surface, and the particle size gradually increases. The large particles will reduce the number of highly-defective grain boundaries and decrease the activity of the electroplated layer. The reason is that the metal deposition rate increases with the increasing current. When the current is less than 0.09 A, the deposition process of Co and Mo keeps stable. According to the above results, the most efficient method to adjust the composition and the CTE of the Co-Mo electroplated layer is to change the content of $Na_2MoO_4$ in the plating solution.

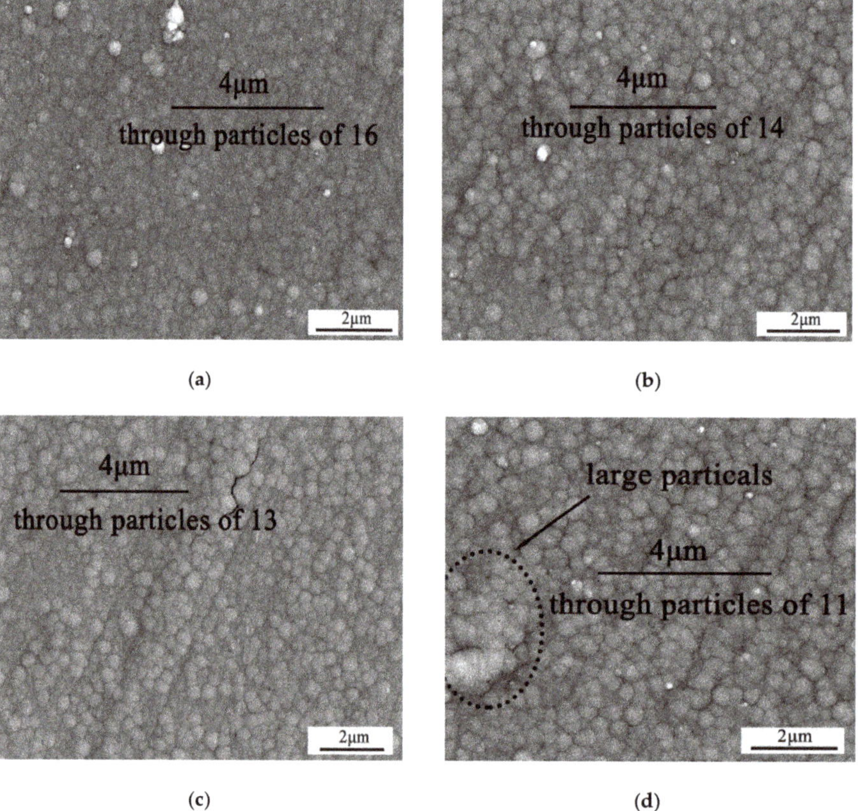


**Figure 3.** The BSE micrographs of the as-electroplated Co-Mo barrier layers with current of: (**a**) 0.09 A, (**b**) 0.11 A, (**c**) 0.13 A and (**d**) 0.15 A. The grain size of the electroplated layer increases gradually with the increasing current.

Figure 4 shows the electroplated layers after annealing at 450 °C for 2 h. For the layers with Mo content greater than 40 wt %, cracks appear in the annealed layer (Figure 4a,b), indicating that the CTE of the layer is less than that of the skutterudite compound. On the other hand, for the layers with Mo content less than 30 wt %, cracks appear in the skutterudite compound (Figure 4d), indicating that the CTE of the layer is greater than that of the skutterudite compound. Figure 4c shows the layer with Mo content of 30 wt %–40 wt %. No cracks are found either in the annealed layer or in the skutterudite compound. We have extended the current method to the *p*-type skutterudite. The preliminary results show that a metallurgical bonding of Co-Mo to the skutterudite compound can also be achieved. Since the CTE of *p*-type skutterudite is larger than that of *n*-type skutterudite, the chemical composition of the Co-Mo layer should be further optimized. The details will be reported in our future manuscripts.

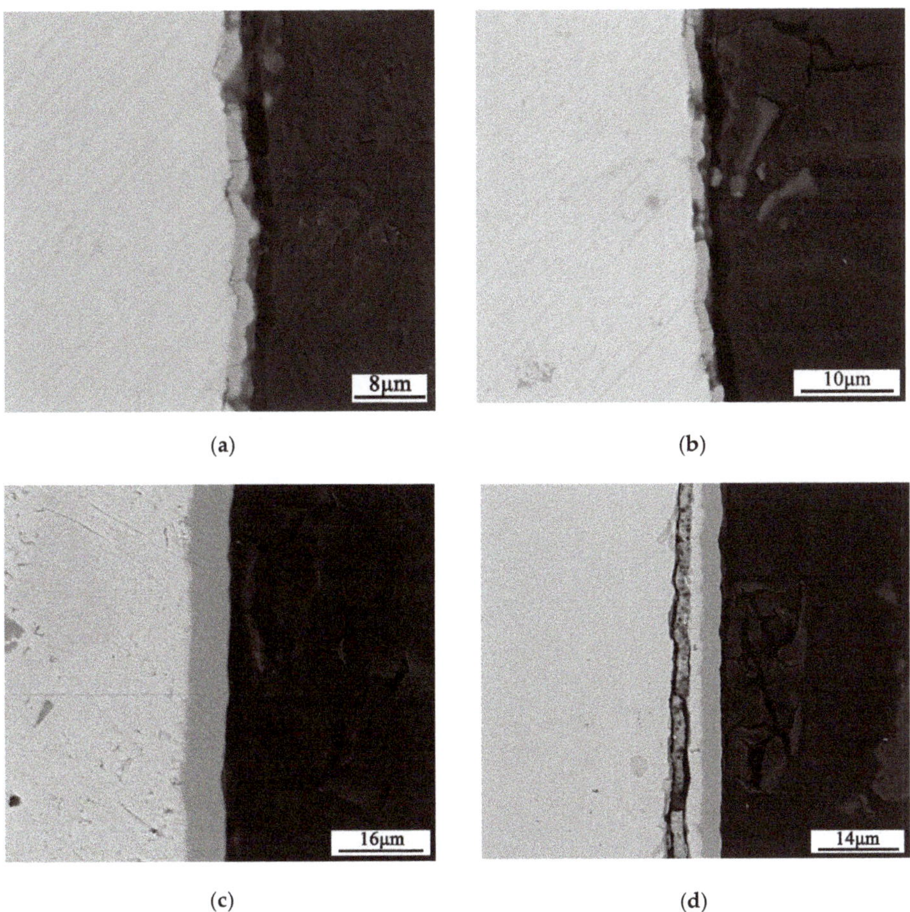


**Figure 4.** The cross-sectional BSE micrograph of the Co-x wt % Mo/SKU interface: (**a**) x = 50, (**b**) x = 40, (**c**) x = 35, (**d**) x = 25. No cracks were found either in the annealed layer or the skutterudite compound when x = 35.

### 3.3. Diffusion Bonding between the Skutterudite Compound and the Cu-Mo Electrode

To join the skutterudite compound to the Cu-Mo electrode, we first prepared a thick Co-Mo electroplated layer on the surface of the skutterudite compound. As shown in Figure 5, the thickness of the layer increases with the increasing electroplating time. The growth rate of the Co-Mo barrier layer is about 2 μm/h.

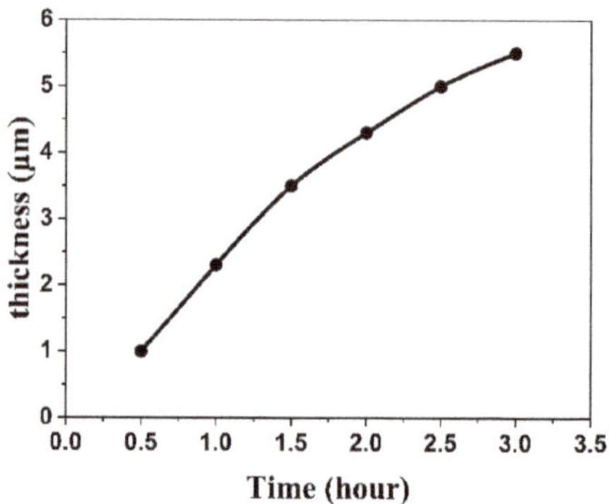

Figure 5. The effects of plating time on the thickness of the electroplated layer.

Figure 6a shows a typical joint of the skutterudite compound metallized by Co-Mo and the Cu-Mo electrode. The joint is fabricated by the diffusion bonding method at 600 °C and 1 MPa for 10 min. A reaction layer, CoSb$_2$ of 2–3 μm, formed between the skutterudite compound and the Co-Mo layer (as shown in Figure 6b). On the contrary, there is no obvious reaction layer between the Co-Mo layer and the Cu-Mo electrode.

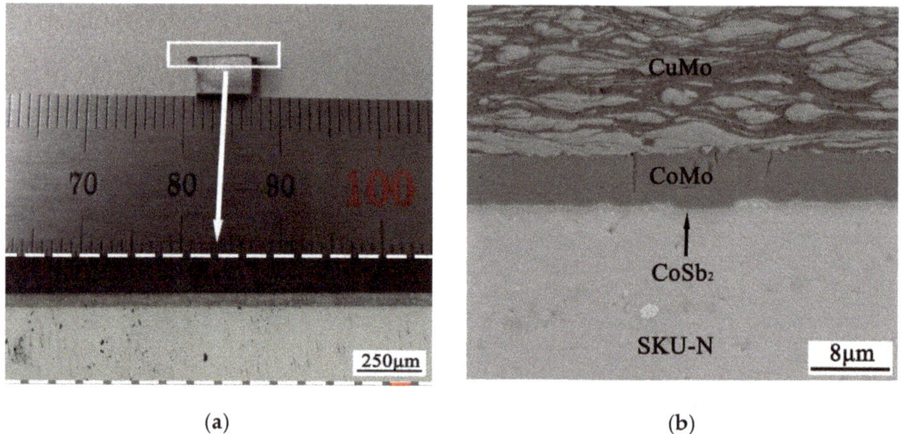


(a)      (b)

**Figure 6.** (**a**) The diffusion bonded joint between the skutterudite compound and the Cu-Mo electrode. (**b**) The BSE image of the cross-section of the joint. A reaction layer (CoSb$_2$) of 2–3 μm forms between the skutterudite compound and the Co-Mo layer after the diffusion bonding process.

To investigate the specific contact resistivity (SCR) of the joint, we also fabricate a skutterudite–electrode–skutterudite sandwich structure (as shown in Figure 7a) using the same technology as described above. The SCR is, thus, measured on this sandwich structure using our homemade instrument. The room-temperature electrical resistivity of the skutterudite alloy is measured by the homemade instrument, too. The value of $2.85 \times 10^{-6}$ Ω m is consistent with the CTA-3 result, indicating the reliability of our homemade equipment.

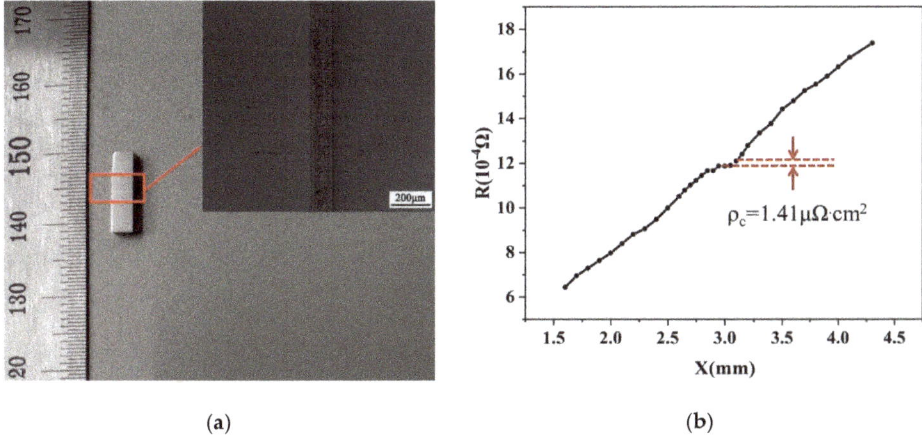

**Figure 7.** (**a**) The skutterudite–electrode–skutterudite sandwich structure fabricated by the diffusion bonding method. (**b**) The measured specific contact resistivity of the Co-Mo/skutterudite interface.

The Co-Mo/SKU interfaces fabricated by the current work exhibit extremely low SCR of 1.41 $\mu\Omega\cdot$cm$^2$ (as shown in Figure 7b). This value is among one of the lowest SCRs to date reported for the thermoelectric modules as shown in Table 2.

**Table 2.** The specific contact resistivity for various classes of thermoelectric materials.

| Thermoelectric Materials | Metallization Layer | Specific Contact Resistivity ($\mu\Omega\cdot$cm$^2$) |
|---|---|---|
| Yb$_{0.3}$Co$_4$Sb$_{12}$ [14] | Mo-Ti | 9 |
| Bi$_2$Te$_3$ [27] | Au | 2.73 |
| Mg$_2$Si [28] | Ni | 210 |
| Hf$_{0.5}$Zr$_{0.5}$CoSn$_{0.2}$Sb$_{0.8}$ [29] | Ag | 38 |
| (Mm,Sm)yCo$_4$Sb$_{12}$ [26] | Fe-Ni | 2 |
| n-type skutterudite [30] | CoSi$_2$ | 0.4 |
| This work | Co-Mo | 1.41 |

Figure 8a shows the microstructure of the CuMo/CoMo/SKU interface after annealing at 550 °C for 60 h. According to the EDS line-scan results (as shown in Figure 8c), the reaction layer mainly contains three parts: the Co-Sb-Mo layer, the CoSb layer and the CoSb$_2$ layer. Compared with the joint before annealing (as shown in Figure 6b), a CoSb layer (~6 μm in thickness) appears between the CoMo and CoSb$_2$ layers. The thickness of the CoSb2 layer increases to about 5 μm. The atomic concentration of Mo increases remarkably in the Co-Sb-Mo layer. Therefore, it can be expected that the enrichment of Mo will suppress the diffusion of Sb by blocking the diffusion path [29]. In fact, Figure 8b shows the evolution of the thickness of the reaction layer of the CoMo/SKU interface as a function of annealing time. It can be seen that the thickness of the diffusion layer is only about 11 μm after annealing at 550 °C for 60 h. As the growth thickness (*L*) of the diffusion layer is linear with the square root of the annealing time (t), it is expressed by the following equation [26]:

$$L = a + b\sqrt{t} \tag{1}$$

where *a* and *b* are fitting parameters. The best fitting shows that $a = 3.7$ μm, $b = 0.92$ μm/h$^{1/2}$. According to the fitting results, the thickness of the diffusion layer after one-year annealing is predicted to be about 90 μm.

Figure 8d shows the measured SCR of the CuMo/CoMo/SKU joint after annealing at 550 °C for 60 h. The SCR of 6.13 μΩ·cm² is slightly higher than that of the as-bonded one. This value is still among the lowest ones according to Table 2. Since the contact resistivity is somehow proportional to the diffusion layer thickness, we can roughly estimate the contact resistivity after one-year annealing is about 50 μΩ·cm². Both the EDS line-scan and the SCR results demonstrate that the metallization layer fabricated in the current work is thermally stable.

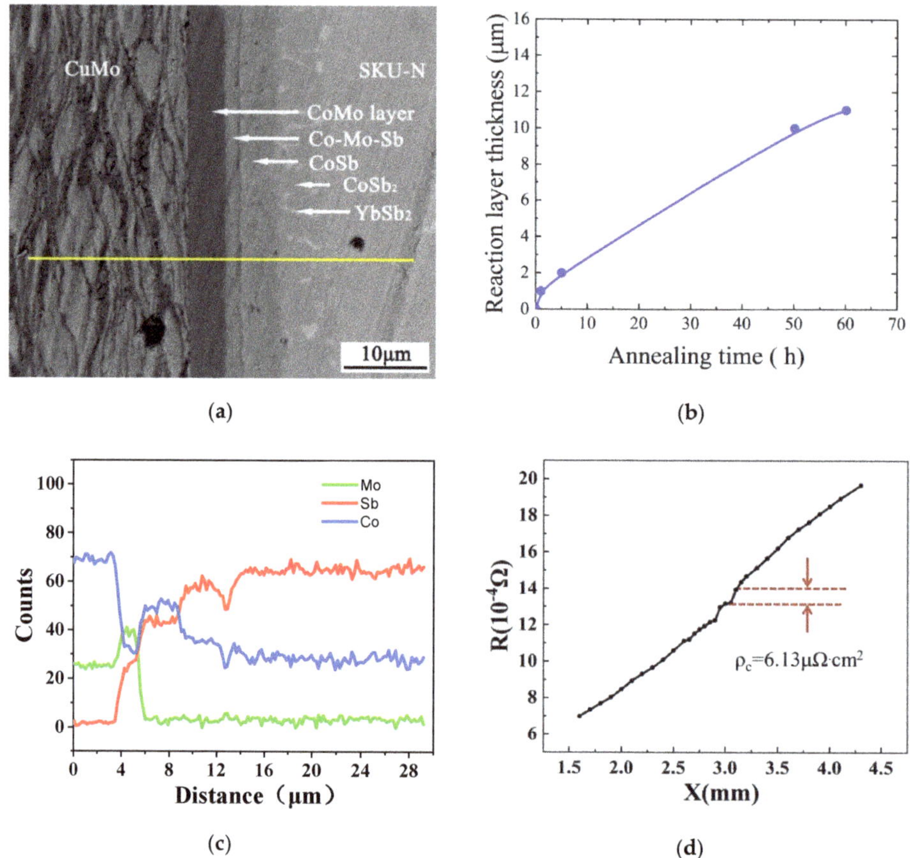


**Figure 8.** (a) The SEM image of the CuMo/CoMo/SKU diffusion bonded joint after annealing at 550 °C for 60 h. About 11 μm reaction layer of Co-Mo-Sb/CoSb/CoSb₂ is formed. (b) The diffusion layer thickness vs. annealing time for the CuMo/CoMo/SKU joint. (c) The EDS line-scan profiles of the CuMo/CoMo/SKU diffusion bonded joint after annealing at 550 °C for 60 h. (d) The specific contact resistivity (SCR) of 6.13 μΩ.cm² of the CuMo/CoMo/SKU joint after annealing at 550 °C for 60 h.

We have tried to measure the shear strength of the joint before and after the long-term annealing. The fractures always appear inside the skutterudite leg instead of the joint itself. This result indicates that the diffusion layer will not weaken the mechanical strength of the joint.

## 4. Conclusions

We introduce an approach to prepare the metallization layer and the subsequent diffusion bonding method for the high-performance multifilled *n*-type skutterudite compounds. Using the electroplating followed by the low-temperature annealing approaches, we successfully fabricated a

Co-Mo metallization layer on the surface of the skutterudite compound. The CTE of the electroplated layer can be optimized by changing the weight ratio between Co and Mo, which is controlled by the electroplating temperature, current and the concentration of the Mo ions in the solution. We also find that the most efficient method to adjust the composition of the Co-Mo layer is to change the concentration of $Na_2MoO_4$ in the plating solution. We then joined the metallized skutterudite leg to the Cu-Mo electrode using a diffusion bonding method performed at 600 °C and 1 MPa for 10 min. The Co-Mo/SKU interfaces exhibit extremely low specific contact resistivity of 1.41 $\mu\Omega\cdot cm^2$. The metallization layer fabricated in the current work is thermally stable after annealing at 550 °C for 60 h.

**Author Contributions:** Conceptualization, L.Z., H.G.; formal analysis, J.Z.; investigation, W.G., S.Z., G.Z.; writing—original draft preparation, H.F.; writing—review and editing, H.G.; supervision, J.F.; funding acquisition, H.G. All authors have read and agreed to the published version of the manuscript.

**Funding:** This research was funded by the National Key Research and Development Program of China, grant number 2017YFB0305700, and the National Natural Science Foundation of China, grant number 51874111.

**Acknowledgments:** We thank Peng He for his kind help in the preparation of the manuscript.

**Conflicts of Interest:** The authors declare no conflicts of interest.

## References

1. Pei, Y.; Shi, X.; LaLonde, A.; Wang, H.; Chen, L.; Snyder, G.J. Convergence of electronic bands for high performance bulk thermoelectrics. *Nature* **2011**, *473*, 66–69. [CrossRef] [PubMed]
2. Snyder, G.J.; Toberer, E.S. Complex thermoelectric materials. *Nat. Mater.* **2008**, *7*, 105–114. [CrossRef] [PubMed]
3. Beretta, D.; Neophytou, N.; Hodges, J.M. Thermoelectrics: From history, a window to the future. *Mater. Sci. Eng. R Rep.* **2019**, *138*, 100501. [CrossRef]
4. Chen, L.; Backhaus-Ricoult, M.; He, L. Fabrication Method for Thermoelectric Device. U.S. Patent 8,198,116, 12 June 2012.
5. Wu, T.; Bai, S.Q.; Shi, X. Enhanced thermoelectric properties of $Ba_xEu_yCo_{(4)}Sb_{(12)}$ with very high filling fraction. *Wuji Cailiao Xuebao J. Inorg. Mater.* **2013**, *28*, 224–228. [CrossRef]
6. Muto, A.; Yang, J.; Poudel, B.; Ren, Z.; Chen, G. Skutterudite unicouple characterization for energy harvesting applications. *Adv. Energy Mater.* **2013**, *3*, 245–251. [CrossRef]
7. Alleno, E.; Lamquembe, N.; Cardoso-Gil, R. A thermoelectric generator based on an n-type clathrate and ap-type skutterudite unicouple. *Phys. Status Solidi a* **2014**, *211*, 1293–1300. [CrossRef]
8. Salvador, J.R.; Cho, J.Y.; Ye, Z.; Moczygemba, J.E.; Thompson, A.J.; Sharp, J.W.; Koenig, J.; Sakamoto, J. Conversion efficiency of skutterudite-based thermoelectric modules. *Phys. Chem. Chem. Phys.* **2014**, *16*, 12510–12520. [CrossRef]
9. Takahashi, K.; Kanno, T.; Sakai, A. Bifunctional thermoelectric tube made of tilted multilayer material as an alternative to standard heat exchangers. *Sci. Rep.* **2013**, *3*, 1501. [CrossRef]
10. Choi, S.; Kurosaki, K.; Li, G. Enhanced thermoelectric properties of Ga and In Co-added $CoSb_3$-based skutterudites with optimized chemical composition and microstructure. *AIP Adv.* **2016**, *6*, 125015. [CrossRef]
11. He, J.; Tritt, T.M. Advances in thermoelectric materials research: Looking back and moving forward. *Science* **2017**, *357*, eaak9997. [CrossRef]
12. Wojciechowski, K.T.; Zybala, R.; Mania, R. High temperature $CoSb_3$-Cu junctions. *Microelectron. Reliab.* **2011**, *51*, 1198–1202. [CrossRef]
13. Park, S.H.; Jin, Y.; Cha, J. High-power-density skutterudite-based thermoelectric modules with ultralow contact resistivity using Fe–Ni metallization layers. *ACS Appl. Energy Mater.* **2018**, *1*, 1603–1611. [CrossRef]
14. Fan, X.C.; Gu, M.; Shi, X. Fabrication and reliability evaluation of Yb0.3Co4Sb12/Mo–Ti/Mo–Cu/Ni thermoelectric joints. *Ceram. Int.* **2015**, *41*, 7590–7595. [CrossRef]
15. Gu, M.; Bai, S.; Xia, X. Study on the high temperature interfacial stability of $Ti/Mo/Yb_{0.3}Co_4Sb_{12}$ thermoelectric joints. *Appl. Sci.* **2017**, *7*, 952. [CrossRef]
16. Guo, J.Q.; Geng, H.Y.; Ochi, T. Development of skutterudite thermoelectric materials and modules. *J. Electron. Mater.* **2012**, *41*, 1036–1042. [CrossRef]

17. Geng, H.; Ochi, T.; Suzuki, S. Thermoelectric properties of multifilled skutterudites with La as the main filler. *J. Electron. Mater.* **2013**, *42*, 1999–2005. [CrossRef]

18. Nie, G.; Li, W.; Guo, J. High performance thermoelectric module through isotype bulk heterojunction engineering of skutterudite materials. *Nano Energy* **2019**, *66*, 104193. [CrossRef]

19. Gu, M.; Xia, X.; Li, X. Microstructural evolution of the interfacial layer in the Ti–Al/Yb$_{0.6}$Co$_4$Sb$_{12}$ thermoelectric joints at high temperature. *J. Alloy. Compd.* **2014**, *610*, 665–670. [CrossRef]

20. Feng, S.; Chang, Y.; Yang, J.; Poudel, B.; Yu, B.; Ren, Z.; Chen, G. Reliable contact fabrication on nanostructured Bi2Te3-based thermoelectric materials. *Phys. Chem. Chem. Phys.* **2013**, *15*, 6757–6762. [CrossRef]

21. Chen, S.; Chu, A.H.; Wong, D.S.H. Interfacial reactions at the joints of CoSb$_3$-based thermoelectric devices. *J. Alloy. Compd.* **2017**, *699*, 448–454. [CrossRef]

22. Song, B.; Lee, S.; Cho, S. The effects of diffusion barrier layers on the microstructural and electrical properties in CoSb$_3$ thermoelectric modules. *J. Alloy. Compd.* **2014**, *617*, 160–162. [CrossRef]

23. Fan, J.; Chen, L.; Bai, S.; Shi, X. Joining of Mo to CoSb$_3$ by spark plasma sintering by inserting a Ti interlayer. *Mater. Lett.* **2004**, *58*, 3876–3878. [CrossRef]

24. Zhao, D.; Li, X.; He, L.; Jiang, W.; Chen, L. Interfacial evolution behavior and reliability evaluation of CoSb$_3$/Ti/Mo-Cu thermoelectric joints during accelerated thermal aging. *Alloy. Compd.* **2009**, *477*, 425–431. [CrossRef]

25. Zhao, D.; Li, X.; He, L.; Jiang, W.; Chen, L. High temperature reliability evaluation of CoSb$_3$/electrode thermoelectric joints. *Intermetallics* **2009**, *17*, 136–141. [CrossRef]

26. Song, J.; Kim, Y.; Cho, B.J. Thermal diffusion barrier metallization based on Co–Mo powder-mixed composites for n-type skutterudite ((Mm, Sm)$_y$Co$_4$Sb$_{12}$) thermoelectric devices. *J. Alloy. Compd.* **2019**, *818*, 152917. [CrossRef]

27. Yong, H.; Na, S.K.; Gang, J.G.; Shin, H.S.; Jeon, S.J.; Hyun, S.; Lee, H.J. Study on the contact resistance of various metals (Au, Ti, and Sb) on Bi−Te and Sb−Te thermoelectric films. *Jpn. J. Appl. Phys.* **2016**, *55*, 06JE03. [CrossRef]

28. Zhang, B.; Zheng, T.; Wang, Q.; Zhu, Y.; Alshareef, H.N.; Kim, M.J.; Gnade, B.E. Contact resistance and stability study for Au, Ti, Hf and Ni contacts on thin-film Mg$_2$Si. *Alloy. Compd.* **2017**, *699*, 1134–1139. [CrossRef]

29. Ngan, P.H.; Van Nong, N.; Hung, L.T. On the challenges of reducing contact resistances in thermoelectric generators based on half-Heusler alloys. *J. Electron. Mater.* **2016**, *45*, 594–601. [CrossRef]

30. Jie, Q.; Ren, Z.; Chen, G. Fabrication of Stable Electrode/Diffusion Barrier Layers for Thermoelectric Filled Skutterudite Devices. U.S. Patent 9,722,164, 2 October 2015.

 *materials*

*Article*

# Structural Properties and Thermoelectric Performance of the Double-Filled Skutterudite $(Sm,Gd)_y(Fe_xNi_{1-x})_4Sb_{12}$


Cristina Artini [1,2,*], Riccardo Carlini [1], Roberto Spotorno [1], Fainan Failamani [3], Takao Mori [3,4] and Paolo Mele [5]

[1] Department of Chemistry and Industrial Chemistry, University of Genova, Via Dodecaneso 31, Genova 16146, Italy

[2] Institute of Condensed Matter Chemistry and Technologies for Energy, National Research Council, CNR-ICMATE, Via De Marini 6, Genova 16146, Italy

[3] National Institute for Materials Science (NIMS), International Center for Materials Nanoarchitectonics (MANA) and Center for Functional Sensor & Actuator (CFSN), Namiki 1-1, Tsukuba 305-0044, Japan

[4] University of Tsukuba, Graduate School of Pure and Applied Sciences, 1-1-1 Tennoudai, Tsukuba 305-8671, Japan

[5] Shibaura Institute of Technology, Omiya Campus, 307 Fukasaku, Minuma-ku, Saitama City, Saitama 337-8570, Japan

* Correspondence: artini@chimica.unige.it



Received: 1 July 2019; Accepted: 30 July 2019; Published: 1 August 2019



**Abstract:** The structural and thermoelectric properties of the filled skutterudite $(Sm,Gd)_y(Fe_xNi_{1-x})_4Sb_{12}$ were investigated and critically compared to the ones in the Sm-containing system with the aim of unravelling the effect of double filling on filling fraction and thermal conductivity. Several samples ($x = 0.50$–$0.90$ and $y = 0.15$–$0.48$) were prepared by melting-sintering, and two of them were densified by spark plasma sintering in order to study their thermoelectric features. The crystallographic study enables the recognition of the role of the filler size in ruling the filling fraction and the compositional location of the $p/n$ crossover: It has been found that the former lowers and the latter moves toward lower $x$ values with the reduction of the filler ionic size, as a consequence of the progressively weaker interaction of the filler with the $Sb_{12}$ cavity. The analysis of thermoelectric properties indicates that, despite the $Sm^{3+}/Gd^{3+}$ small mass difference, the contemporary presence of these ions in the $2a$ site significantly affects the thermal conductivity of both $p$- and $n$-compositions. This occurs by reducing its value with respect to the Sm-filled compound at each temperature considered, and making the overall thermoelectric performance of the system comparable to several multi-filled (Fe, Ni)-based skutterudites described in the literature.



**Keywords:** thermoelectricity; skutterudites; crystal structure; powder x-ray diffraction; thermal conductivity


---

## 1. Introduction

In the field of modern solid state chemistry and physics, thermoelectricity occupies a central role due to its relevance in the framework of energy conversion, in particular, refrigeration and electric power generation. Moreover, energy harvesting to power myriad sensors and devices for the Internet of Things (IoT) society is also a promising application [1,2]. The development of thermoelectric technology exploitable for large-scale employments implies in the first instance the search for materials fulfilling several requirements, such as high electrical conductivity and low thermal conductivity. The dimensionless thermoelectric figure of merit $ZT$ can be in fact expressed as follows:

$$ZT = \frac{\sigma S^2}{\lambda_e + \lambda_{ph} + \lambda_{bp}} \tag{1}$$

where $T$ is the absolute temperature, $\sigma$ the electrical conductivity, $S$ the Seebeck coefficient, $\lambda_{el}$, $\lambda_{ph}$ and $\lambda_{bp}$ the electron, phonon and bipolar [3,4] contributions to thermal conductivity, with $\lambda_{ph}$ being the prevailing one in semiconductors, and $\lambda_{bp}$ appearing only above a certain temperature threshold which depends on the system considered. Electrical conductivity and electron thermal conductivity are correlated through the Wiedemann-Franz law, and for this reason they cannot be separately manipulated. The need for decoupling thermal and electrical conductivity suggests that a two-fold strategy is pursued, consisting of mainly acting on the reduction of $\lambda_{ph}$ and the increase of $\sigma$. This goal can be achieved by addressing the search for valid thermoelectric materials for semiconductors. In semiconductors, in fact, at variance with metals, $\sigma$ increases with increasing temperature, and the quantity $\sigma S^2$ (also called the power factor) presents a maximum within their carrier density range. Moreover, $\lambda_{ph}$ prevails over $\lambda_e$, making it possible to separately act on the two terms.

While the improvement of the power factor needs efforts in band structure engineering in order to optimize the carrier concentration [5,6], the manipulation of $\lambda_{ph}$ has been of the focus researchers for a long time, since this parameter is considered the easiest one to be addressed by a phenomenological approach. $\lambda_{ph}$ can be described by the following Equation:

$$\lambda_{ph} = \frac{1}{3}C_v l \tag{2}$$

where $C_v$ is the specific heat at a constant volume, $l$ is the mean free path of phonons, and $v$ is the sound velocity. The reduction of $l$, leading to the reduction of $\lambda_{ph}$, can be achieved through the introduction of scattering centers into the lattice which are able to interfere with phonon transmission, such as the point and volume defects (by alloying) [7,8], grain boundaries (by nanostructuring) [9], nanoinclusions [10,11] secondary phases [12,13], vacancies causing mass fluctuations [14], nano- and micro-sized pores [15,16], or loosely bound atoms.

The last mentioned issue, mainly ruled by the phonon glass electron crystal (PGEC) theory [17], opened the way to a widespread investigation of intermetallic compounds characterized by a high electrical conductivity coupled to a crystal structure provided with large voids, and able to host foreign atoms of proper size. According to the basic concept of the aforementioned theory, in fact, such materials are expected to exhibit a dramatic decrease of phonon thermal conductivity due to the rattling movement of host ions within the cage, without affecting the mobility of charge carriers. These features are actually displayed by members of several classes of intermetallic compounds, such as skutterudites [18], clathrates [19,20], and half-Heusler phases [21]. Skutterudites $MX_3$ [18,22] (where M is a transition metal, such as Co, Fe, Rh or Ir and X a pnicogen atom), in particular, exhibit a body-centered cubic cell (Pearson symbol $cI32$, $Im\bar{3}$ space group, isotypic crystal: $CoAs_3$) presenting two distinct atomic sites: the $8c$ ($\frac{1}{4}$, $\frac{1}{4}$, $\frac{1}{4}$) and the $24g$ ($0$, $y$, $z$), occupied by M and X, respectively. Consequently, M forms with X a strongly tilted corner-sharing octahedra, and a $X_{12}$ icosahedral cage with its center in the $2a$ site located in ($0, 0, 0$) appears. The occurrence of the cited cavity is of primary importance for the thermoelectric performance of the material [23]. The skutterudite itself, in fact, is characterized by an excessively high value of thermal conductivity to be reasonably employed for thermoelectric applications [24]. On the contrary, when an atom of proper size is placed into the void, $\lambda_{ph}$—and hence thermal conductivity—is strongly lowered due to the vibrational modes of the rattling guest atom, which disturb the propagation of heat-carrying phonons, as experimentally observed [25]. Clearly, the size of the guest atom must fit the cavity, namely it has to be large enough to be retained within the void, but at the same time small enough to properly vibrate around its equilibrium position. Ideally, in order to maximize the effect of the $\lambda_{ph}$ reduction, even constraints on the atomic mass of the filler should be obeyed. Generally speaking, a large atomic mass is preferable in order to favor low frequency vibrational modes. Considering both the size and the mass requirements, the most suitable atoms are lanthanides and alkaline-earth elements [18,22,26].

It is noteworthy that, in addition to the $\lambda_{ph}$ reduction, the introduction of the filler ion also exerts a significant effect on the electronic properties of the material. While in fact the skutterudite $CoSb_3$ is a compensated semiconductor, the substitution of Co by the mixture Fe/Ni causes an electronic imbalance, which is only partly restored by the insertion into the structure of the filler ion. It is generally accepted that a larger or smaller filler amount than the one corresponding to the structural and electronic stability of the resulting compound can be hardly forced into the cavity [18]. The dependence of the Yb solubility in $CoSb_3$ on the thermal treatment temperature, for instance, has been recently discussed [27]. However, a large number of studies demonstrated that the filling fraction of the voids depends in the first instance on the oxidation state of the filler, and secondly, on its ionic size, and therefore, it cannot be ad libitum varied. In the $RE_y(Fe_xNi_{1-x})_4Sb_{12}$ systems (RE $\equiv$ rare earth ion), for example, the filling fraction by trivalent lanthanide ions follows a linear trend as a function of $x$, with different fillers presenting at least roughly comparable amounts at a given Fe/Ni content, and small differences attributable to their different ionic sizes [28–35].

An even more effective approach aimed at the depression of $\lambda_{ph}$ consists of the introduction of a structural disorder through the substitution of Co by Fe/Co [36–39], Co/Ni [40,41] Fe/Ni [29,34,38,42,43] or Fe/Co/Ni [30,31], the substitution of Sb by Sb/Sn or Sb/Ge [36], or multiple filling of the cavity by DD (didymium) [36,38,41], Mm (mischmetal) [31,41], Ce/Yb [30], Ce/Nd [44], Ce/Yb/In [45] or by a mixture of alkaline-earths [46] with the possible addition of lanthanide elements [41,42] and elements of the IIIB group [47]. Multiple filling, in particular, was thoroughly studied [48], since it ensures a broadening of the range of phonon frequencies which can be perturbed by the rattling movement of doping ions. For this reason, when designing a multiple doping, a significant difference in the mass of the different fillers is generally recommended.

In recent years, the synthetic procedure [49], the room [34] and high temperature [35] structural properties, as well as the thermoelectric [50,51], thermal [52] and mechanical [53] features, and the corrosion behavior [54] of the $Sm_y(Fe_xNi_{1-x})_4Sb_{12}$ system, were extensively studied by the present research group. In this work, the addition of a second filler element has been attempted in order to lower thermal conductivity and hence to improve ZT. Gd was chosen taking into account its larger mass and slightly smaller size with respect to Sm. The structural properties of the obtained double-filled skutterudite are analyzed relying on powder x-ray diffraction data mainly in comparison to the only Sm-doped compound. The thermoelectric performance is discussed in the light of literature data collected on multiple-doped (Fe,Ni)-based skutterudites.

## 2. Materials and Methods

### 2.1. Synthesis

Five compositions belonging to the $(Sm,Gd)_y(Fe_xNi_{1-x})_4Sb_{12}$ system, with nominal $x = 0.90, 0.80, 0.70, 0.63$ and $0.50$, were synthesized by the direct reaction of pure elements Fe (Alfa-Aesar, 99.99 wt %), Ni, Sm, Gd (NewMet, 99.9 wt %) and Sb (Mateck, 99.999 wt %). The compositions were chosen in order to have $p$- ($x = 0.90, 0.80, 0.70$) and $n$-type ($x = 0.50$) samples, as well as a composition located in the $p/n$ crossover region ($x = 0.63$). As aforementioned, the filler content is strictly related to the electronic count of the compound, and hence to the Fe/Ni ratio. For this reason, the total rare earth amount to be used ($y$) was determined relying on the results of a previous study performed on the $Sm_y(Fe_xNi_{1-x})_4Sb_{12}$ system [34]. Moreover, considering the larger size of $Sm^{3+}$ with respect to $Gd^{3+}$, which is expected to allow a better accommodation of the former within the cavity, the 2:1 ratio for the Sm:Gd amount was chosen. A slight Sb excess with respect to the stoichiometric amount was used due to its non-negligible vapor pressure ($10^{-3}$ mm Hg at 873 K [55]). The elements mixture was placed into an Ar-filled quartz ampoule, which was then sealed under an Ar flow, thermally treated at 1223 K for

3 h, and subsequently rapidly cooled in an iced water bath. The obtained samples were then annealed at 873 K for 4 days.

Afterwards, a *p*- and an *n*- composition, namely samples with nominal $x = 0.80$ and $0.50$, were densified in order to obtain suitable specimens to be submitted to the evaluation of the thermoelectric performance. To this purpose, both specimens were first ball milled at a rotation speed of 150 rpm for 1 h employing a steel jar and steel balls. Subsequently, they were submitted to spark plasma sintering (SPS) by applying in vacuum ($P = 5 \times 10^{-2}$ atm) a pressure of 50 MPa at 773 K for 20 min. The discs having 10 mm diameter and 2 mm thickness were obtained. The density of the obtained samples was calculated from their mass and dimension.

The samples were named Fe50_bulk, Fe50_SPS, and so on, according to the % Fe amount with respect to the total (Fe + Ni) content, and to the treatment they were submitted to (annealing: _bulk; annealing followed by SPS: _SPS).

### 2.2. Optical and Electronic Microscopy

The morphology, microstructure, porosity and composition were evaluated by analyzing micrographically polished surfaces of all the samples through optical and electron microscopy. In particular, the specimens were observed by an optical microscope (OM, Leica MEF4M, Leica, Wetzlar, Germany) equipped with a computerized image capture and processing software (Zeiss Axiovision 4, Zeiss, Oberkochen, Germany). A Zeiss SUPRA 40 VP-30-51 scanning electron microscope (Zeiss, Oberkochen, Germany) with a field emission gun (FE-SEM), equipped with a high sensitivity 'InLens' secondary electron detector and with an EDS microanalysis INCA Suite Version 4.09 (Oxford Instruments, Abingdon-on-Thames, UK), was employed to study the morphology and evaluate the composition of each phase. The EDS data were optimized using Co as a standard. The microphotographs were taken on each sample both by backscattered and secondary electrons, and the EDS analyses were performed on at least five points (acquisition time: 60 s; working distance: 8.5 mm). The porosity of the bulk and SPS samples was evaluated by image analysis applied both to optical and electron microscopy microphotographs using the Fiji-ImageJ, version 1.49b [56,57].

### 2.3. X-ray Diffraction

All the bulk and SPS samples were powdered, sieved through a 44 µm sieve, placed on a Si zero-background sample holder, and subsequently analyzed by a Bragg-Brentano powder diffractometer (Philips PW1050/81, Fe-filtered Co $K_\alpha$ radiation) (Philips, Amsterdam, Netherlands). The diffraction patterns were collected in the angular range 15°–120°, with angular step 0.02°, and counting time 17 s. The structural model of all the occurring phases was refined by the Rietveld method using the FullProf software [58]. The accurate lattice parameters of the skutterudite were obtained from diffractograms collected making use of Ge as an internal standard.

### 2.4. Microhardness

The microhardness of the samples Fe50_bulk, Fe50_SPS, Fe80_bulk and Fe80_SPS was evaluated by a Leica VMHT30A microhardness tester (Leica, Wetzlar, Germany) provided with Vickers indenter. The indentations were done on polished and unetched samples at random positions. Further, 20 tests were performed on each sample applying a test load of 50 g for 15 s. The errors associated with the microhardness average values corresponded to the standard deviation.

### 2.5. Measurements of Transport Properties

The thermal diffusivity ($\alpha$) and specific heat ($C_p$) were measured by a laser flash method using a Netzsch Hyperflash 467 under flowing $N_2$ gas (Netzsch, Selb, Germany). A pyroceram standard was used as the specific heat reference. Prior to the measurements, the samples and standard were evenly coated with a thin layer of graphite. For each temperature, 5 data points were obtained and averaged. The SPS samples, with diameter of ~10 mm and thickness of ~2 mm were used directly, while bulk samples were cut into ~6 × 6 × 2 mm$^3$ square plates for the measurement. Thermal conductivity ($\lambda$) was calculated from the measured thermal diffusivity, specific heat, and density ($d$) using the relationship:

$$\lambda = \alpha C_p d \tag{3}$$

The cuboids with dimensions of ~2 × 2 × 8 mm$^3$ were cut from both SPS and bulk samples and used for the Seebeck and electrical resistivity measurements. The simultaneous measurements of Seebeck and the electrical resistivity were performed on a ZEM-2 apparatus (ULVAC-Riko, Chigasaki, Japan) under a low pressure of He atmosphere. The average relative uncertainties of the measurements were 6%, 8%, 11% and 19% for the Seebeck coefficient, electrical resistivity, thermal conductivity, and ZT, respectively [59].

## 3. Results and Discussion

### 3.1. Compositional and Morphological Characterization

According to SEM microphotographs, the EDS analyses and X-ray diffraction, the filled skutterudite results were by far the main phase for each formulation, even if small amounts of extra phases could be revealed in all the samples.

In Table 1, the experimental composition of each specimen, as obtained from Rietveld refinements, is reported. Due to the similarity in the X-ray scattering factor of Fe and Ni, the Fe/Ni ratio of skutterudite was not refined, but fixed to the value provided by the EDS measurements. The Sm and Gd amounts derive from the refinement of the occupancy factor of the 2a site, and they are in good agreement with the EDS results. It can be observed that in all the samples, the experimental Fe and Ni content were very close to the nominal values. The third column of the Table collects a list of the additional phases identified for each composition. An estimate of the weight percent is reported only when the extra phase could be refined and their amount evaluated by the Rietveld method. When, on the contrary, the presence of the phase could only be revealed by the EDS, but not by X-ray diffraction, its amount is defined as traces. It can be noticed that the most commonly occurring additional phases were Sb and the binary compound (Fe,Ni)Sb$_2$, and to a minor extent (Gd,Sm)Sb$_3$ and the ternary compound (Ni,Fe)(Gd,Sm)Sb$_3$. Among the agreement factors of the Rietveld refinements, R$_B$ of skutterudite was quite low, so confirming the accuracy of the structural model. The relatively high values of $\chi^2$ for samples Fe70_bulk and Fe50_bulk are attributable to the very low amount of the secondary phase (Fe,Ni)Sb$_2$, which does not allow to correctly refine the model, in particular the Fe/Ni ratio. Regarding Fe80_SPS, the significant decrease of Sb with respect to Fe80_bulk, accompanied by the increase in (Fe,Ni)Sb$_2$ and occurrence of traces of (Sm,Gd)$_2$Sb$_5$ (see Table 1), could result from the effect of vacuum environment during SPS process. The considerably high vapor pressure of Sb at the sintering temperature, coupled to the vacuum condition, could have promoted the reaction of the free Sb present in Fe80_bulk with Fe and Ni deriving from skutterudite. The albeit limited presence of (Gd,Sm)-containing extra phases allows the conclusion that the rare earth content of the filled skutterudite is an intrinsic property of each composition, which is independent of the availability of the filler ions. Figure 1 shows, as a representative example, a microphotograph taken by backscattered electrons on the sample Fe90_bulk, where the presence of the skutterudite-based matrix and small amounts of extra phases can be clearly identified. The figure, moreover, allows the recognition that the average skutterudite crystal size ranges between 5 and 10 μm.

**Table 1.** The refined compositions, list of additional phases, and agreement factors of the Rietveld refinements ($\chi^2$ and skutterudite $R_B$).

| Sample | Refined Composition | Additional Phases | $\chi^2$ | $R_B$ |
|---|---|---|---|---|
| Fe90_bulk | $Sm_{0.39(1)}Gd_{0.08(1)}(Fe_{0.87(1)}Ni_{0.13(1)})_4Sb_{12}$ | Sb (1 wt.%), $(Fe,Ni)Sb_2$ (14 wt.%), $(Fe,Ni)(Gd,Sm)Sb_3$ * | 6.4 | 5.8 |
| Fe80_ bulk | $Sm_{0.31(1)}Gd_{0.11(1)}(Fe_{0.78(1)}Ni_{0.22(1)})_4Sb_{12}$ | Sb (11 wt.%), $(Fe,Ni)Sb_2$*, $(Sm,Gd)_2Sb_5$ * | 4.0 | 4.0 |
| Fe80_SPS | $Sm_{0.34(1)}Gd_{0.08(1)}(Fe_{0.76(1)}Ni_{0.24(1)})_4Sb_{12}$ | Sb (2 wt.%), $(Fe,Ni)Sb_2$ (7 wt.%), $(Sm,Gd)_2Sb_5$ * | 4.2 | 3.0 |
| Fe70_ bulk | $Sm_{0.23(1)}Gd_{0.08(1)}(Fe_{0.68(1)}Ni_{0.32(1)})_4Sb_{12}$ | Sb (11 wt.%), $(Fe,Ni)Sb_2$*, $Sb_5(Gd,Sm)_2$ * | 14.8 | 6.9 |
| Fe63_ bulk | $Sm_{0.19(1)}Gd_{0.08(1)}(Fe_{0.63(1)}Ni_{0.37(1)})_4Sb_{12}$ | Sb *, $(Fe,Ni)SmSb_3$ * | 5.8 | 4.8 |
| Fe50_ bulk | $Sm_{0.09(1)}Gd_{0.05(1)}(Fe_{0.51(1)}Ni_{0.49(2)})_4Sb_{12}$ | $(Fe,Ni)Sb_2$ (1 wt.%), $(Ni,Fe)(Gd,Sm)Sb_3$ * | 12.2 | 4.7 |
| Fe50_SPS | $Sm_{0.08(1)}Gd_{0.07(1)}(Fe_{0.51(1)}Ni_{0.49(2)})_4Sb_{12}$ | $(Fe,Ni)Sb_2$ (2 wt.%), $(Ni,Fe)(Gd,Sm)Sb_3$ *, $Ni_{0.4}Sb_2(Sm,Gd)$ * | 6.5 | 5.8 |

* traces.

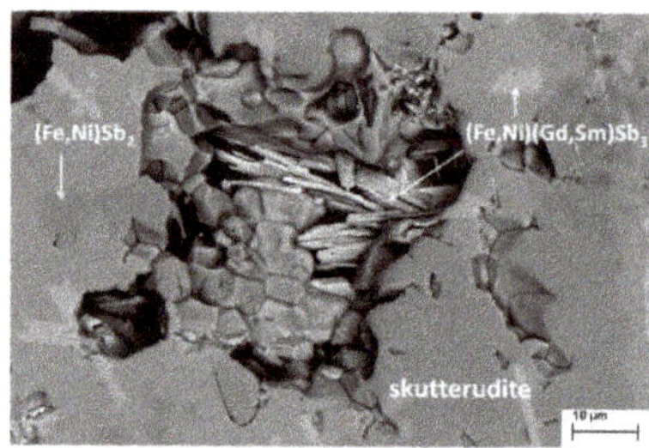


**Figure 1.** SEM microphotograph taken by backscattered electrons on the polished surface of sample Fe90_bulk.

### 3.2. Structural Features

Each X-ray powder diffraction pattern was treated by refining the structural model of the skutterudite. To this purpose, the previously described cubic cell crystallizing in the $Im\bar{3}$ space group was considered. When possible, even the structural models of the additional phases Sb and $(Fe,Ni)Sb_2$ were optimized. For every diffractogram, ~70 experimental points chosen from the collected pattern were fitted in order to model the background, while the peak profiles were optimized by means of the pseudo-Voigt function. In the last refinement cycles, the structural parameters of the skutterudite (i.e. the cell parameter, the $x$ and $y$ atomic coordinates of Sm, and the Sm and Sb occupation in the $2a$ and $24g$ positions, respectively), as well as its scale factor, nine peak parameters, and the background points, were refined at the same time. The individual isotropic displacement parameters $B_{iso}$ were optimized for each atom while keeping the fixed atomic occupancy factors. As aforementioned, Fe and Ni occupancy factors were fixed to the value deriving from the EDS analysis and were not allowed to vary. Figure 2 represents the Rietveld refinement plot of sample Fe63_bulk. The agreement factors are collected in Table 1.

The trend of the lattice parameter as a function of the Fe content ($x$ in $(Sm,Gd)_y(Fe_xNi_{1-x})_4Sb_{12}$) is reported in Figure 3. The presence in the same graph of the experimental values obtained from $Sm_y(Fe_xNi_{1-x})_4Sb_{12}$ [34] allows for an immediate comparison between the two systems and a direct analysis of the effect of Gd addition. Firstly, it can be observed that the lattice parameter grows with increasing the Fe value, owing to the larger size of the latter with respect to Ni, and also to the increasing amount of the filler ion hosted within the $Sb_{12}$ cage with increasing $x$. Moreover, similarly to what revealed in the Sm-containing skutterudite, even if in the presence of a limited number of compositions, the trend of the experimental data suggests the existence of a slope change occurring close to the $p/n$ crossover, possibly attributable to the $Fe^{2+}$ high- to low-spin transition occurring with increasing $x$, as widely discussed in [34]. On the whole, it can be concluded that no significant differences in the lattice parameter can be appreciated between the samples belonging to both series.

Nevertheless, the non negligible difference in the ionic size of $Sm^{3+}$ and $Gd^{3+}$ ($r_{Sm3+}$ [CN12] = 1.24 Å; $r_{Gd3+}$ [CN9] = 1.107 Å) [60] allows the hypothesis that the Gd filler ion could be less strongly retained within the $Sb_{12}$ cage than Sm, with what is in principle expected to be responsible for a lower filling degree of the void in (Sm,Gd) double-filled samples.

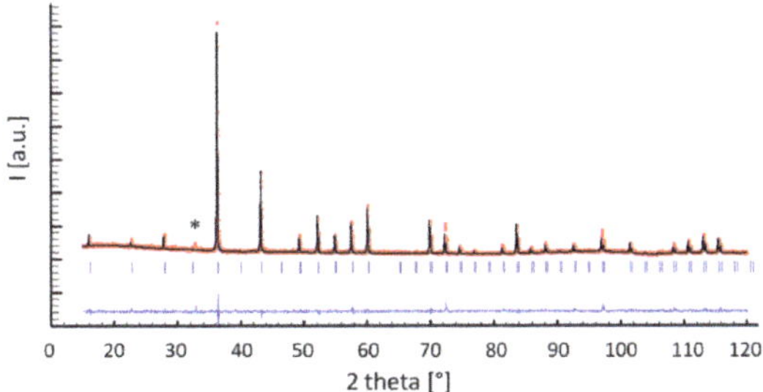


**Figure 2.** The Rietveld refinement plot of the sample Fe63_bulk. The red and black lines are the experimental and the calculated diffractogram, respectively; the lower line is the difference curve; the blue vertical bars indicate the calculated positions of Bragg peaks of the filled skutterudite; the asterisk (*) indicates the position of the main peak of Sb.

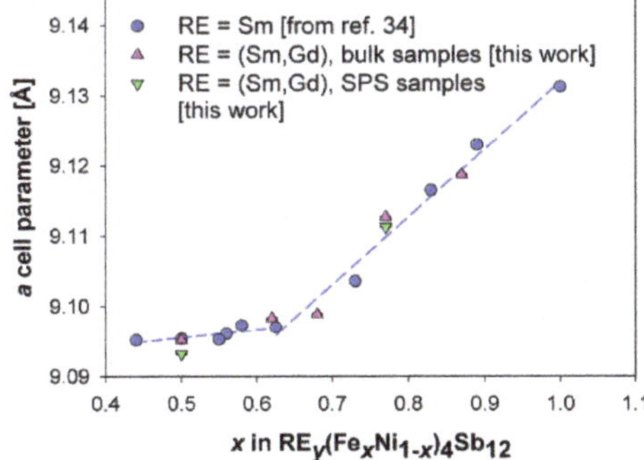


**Figure 3.** The trend of the lattice parameter of the bulk and SPS samples belonging to the $(Sm,Gd)_y(Fe_xNi_{1-x})_4Sb_{12}$ as a function of the Fe amount, compared to the values obtained from the $Sm_y(Fe_xNi_{1-x})_4Sb_{12}$ system (taken from [34]). The dashed lines are regression lines fitting experimental data deriving from [34]. The error bars are hidden by data markers.

With this in mind, the diagram appearing in Figure 4 has been built, showing the refined values of the total lanthanide content as a function of the Fe content [$y$ vs. $x$ in $(Sm,Gd)_y(Fe_xNi_{1-x})_4Sb_{12}$]. The refined values of the $2a$ site occupancy factor obtained from the only Sm- [34] and Ce-containing system [29,30,61] are reported too as terms for comparison. It has to be underlined that the cited representation is of primary importance for the investigation of the skutterudite properties, since it is able to act as a bridge between the structural and electronic features of the material. As aforementioned, in fact, the filler ion provides a certain number of electrons—according to its oxidation state and amount—which determines the filling fraction of the cavity, expressed in Figure 4 as the occupancy factor of the $2a$ site. The black line also appearing in the figure represents for each x value the filling fraction of the cavity which would ensure the electronic count of a compensated semiconductor. The crossing point between the regression line fitting experimental data and the aforementioned black line, represents the $p/n$ crossover: At $x$ higher than the one corresponding to the crossover, the experimental points lie below the black line, meaning that—at least relying on structural data—a $p$-type conduction regime occurs. The opposite happens at $x$ lower than the crossing point. The measurements of the room temperature

Seebeck coefficient nicely confirm the accuracy of this determination for $Sm_y(Fe_xNi_{1-x})_4Sb_{12}$ [34] and $Ce_y(Fe_xNi_{1-x})_4Sb_{12}$ [29].

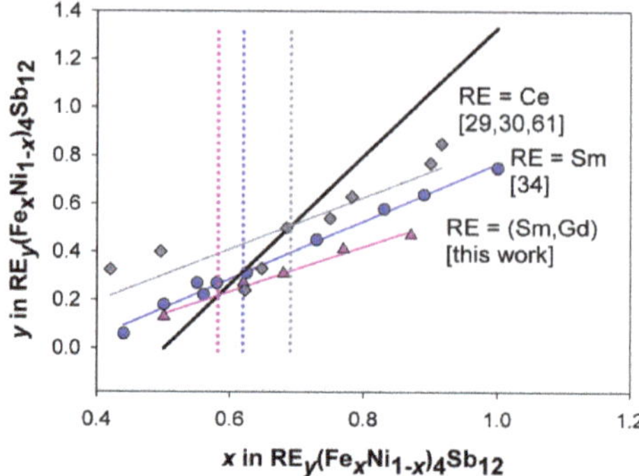


**Figure 4.** The rare earth content as a function of the Fe amount in different (Fe-Ni)-based skutterudites. The black thick line represents the theoretical amount of trivalent rare earth necessary to reproduce the electronic count of the compensated semiconductor $CoSb_3$; experimental values (deriving from this work and from references [29,30,34,61] are fitted by regression lines. The dotted vertical lines pinpoint the position of the *p/n* crossover for the different systems.

Having said that, the analysis of Figure 4 indicates that the (Sm,Gd)-based system presents for each x a rare earth amount significantly lower than the Sm- and the Ce-containing one, leading to roughly parallel regression lines for the three series. The reciprocal position of the regression line with respect to the one of the compensated semiconductor suggests that in the doubly filled system, the *p/n* crossover takes place at the lowest Fe amount, namely at $x\sim0.59$, being $x\sim0.63$ and $x\sim0.70$ the crossover points for $Sm_y(Fe_xNi_{1-x})_4Sb_{12}$ and $Ce_y(Fe_xNi_{1-x})_4Sb_{12}$, respectively. These informations are indeed of practical relevance, since they have to be taken into account when designing the thermoelectric device. However, they are even more interesting when considering the viewpoint of basic science, in particular when the question to be answered is: Which contributions drive the filling degree of a filled skutterudite?

As aforementioned in the Introduction, it is difficult to force into the structure more filler ions than needed to provide the electronic stability to the resulting skutterudite. Therefore, in different $RE_y(Fe_xNi_{1-x})_4Sb_{12}$ systems, a rough similarity among the content of fillers in the same oxidation state can be observed at a given x value, particularly in the compositional region close to the *p/n* crossover [28–34]. On the contrary, significantly different values are observed when alkaline earth ions are inserted, since the $2^+$ oxidation state results into a higher filler amount allowed into the structure [28]. Nonetheless, even if the main driving force ruling the filling fraction is the filler oxidation state, the size effect cannot be neglected. The data reported in Figure 4 clearly show that the filling fraction decreases at each x value following the sequence Ce→Sm→(Sm,Gd), i.e. with decreasing the filler ionic size. According to Shannon [60], in fact, ionic radii are 1.34 Å (CN12), 1.24 Å (CN12) and 1.107 Å (CN9) for $Ce^{3+}$, $Sm^{3+}$ and $Gd^{3+}$, respectively. The described evidence thus suggests that the larger the ions, the stronger their interaction with the cavity, and consequently the higher their content. Moreover, this effect forces the *p/n* crossover to shift toward higher x values with increasing the ionic size of the filler. In view of the above, the low filler content observed at each x value in the (Sm,Gd)-based system can be attributed to the small size of Gd, which makes the (Sm,Gd) mixture on the whole the least retained within the void among the systems considered. This hypothesis can be also discussed in the light of the trend of the Sm/(Sm + Gd) amount ratio as a function of x, reported in Figure 5. The almost regular increase in the aforementioned amount ratio with increasing the Fe content, and hence the cavity size [34], is due to the progressive growth of the Sm content, accompanied by the substantially constant amount of Gd, as can be inferred from the refined compositions reported in Table 1. In particular, considering that during synthesis Sm and Gd were added in the 2:1 ratio, the analysis of Figure 5 suggests that the experimental rare earths ratio starts deviating from the nominal value already for $x = 0.63$, where it results equal to 0.7, and it reaches 0.83 for $x = 0.90$. The contribution of the filler ion size to the filling fraction and to the position of the $p/n$ crossover is therefore further confirmed.

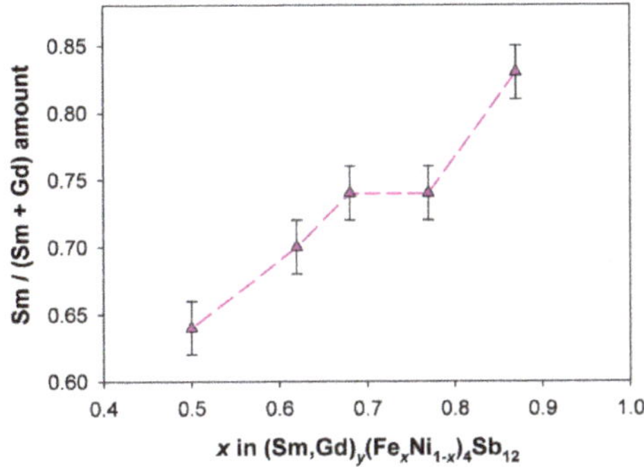

**Figure 5.** The trend of the refined Sm/(Sm + Gd) amount ratio as a function of the Fe content.

Relying on the present results, it can be thus concluded that multi-filling involving the addition of a small rare earth filler (i.e., a filler which significantly lowers the filling fraction) is expected to exert a two-fold effect. On one hand, it leads to a more effective phonon scattering due to its large atomic mass and small ionic radius, which bring about low frequency modes well interacting with phonons responsible for heat flow in solids. On the other hand, such small ions are weakly bound to the $Sb_{12}$ cage, so that their filling degree is lower than expected for larger, isovalent ions. The latter factor is in principle detrimental, as a large filling degree is desirable, being a regular decrease in the phonon contribution to thermal conductivity generally observed with increasing the filling fraction [62]. Therefore, the choice of filler ions and their proportion is a crucial point in designing multi-filled skutterudites, and it needs to be carefully operated, in order to finely tune the contribution to phonon scattering of fillers with different sizes, keeping in mind the final goal of minimizing thermal conductivity.

*3.3. Densification of Samples*

The effect of SPS on the samples Fe80 and Fe50 was evaluated by considering porosity, microhardness and density of the samples before and after the treatment. In Figures 6 and 7, the microphotographs taken by the secondary electrons prior to and after SPS on the polished surface of the samples Fe80 and Fe50, respectively, are shown. A substantial reduction of porosity can be clearly observed, as also confirmed by the results of the image analysis reported in Table 2, together with a significant density increase. Even the results of the Vickers microhardness measurements point at a hardness increase due to the SPS treatment, as a consequence of the porosity reduction, which limits the probability of including holes within indentations.

**Table 2.** The porosity, Vickers microhardness, the absolute and relative density of the samples Fe50 and Fe80 before and after the SPS treatment.

| Sample | Porosity [%] | Vickers Microhardness | Density [g/cm$^3$] | Density [%] |
|---|---|---|---|---|
| **Fe80_bulk** | 19 | 328(70) | 6.40 | 82 |
| **Fe80_SPS** | 10 | 433(50) | 6.87 | 88 |
| **Fe50_bulk** | 16 | 460(49) | 6.62 | 87 |
| **Fe50_SPS** | 4 | 475(55) | 7.33 | 97 |

(a)                    (b)

**Figure 6.** SEM microphotograph taken by secondary electrons on the polished surface of sample (**a**) Fe80_bulk and (**b**) Fe80_SPS.

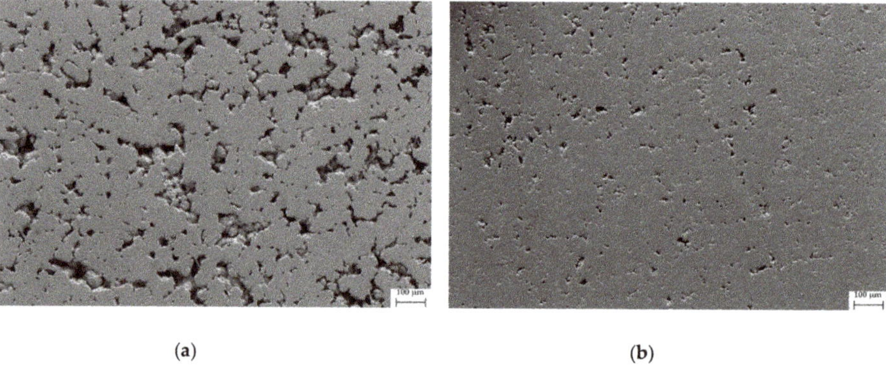

(a)                    (b)

**Figure 7.** SEM microphotograph taken by secondary electrons on the polished surface of sample (**a**) Fe50_bulk and (**b**) Fe50_SPS.

The application of an external pressure is possibly expected to reflect also on the structural, microstructural and compositional features of a material. Focusing on skutterudites, different densification methods resulted to affect the lattice parameter [42,63], filling fraction [64], peak width [42], and the amount of additional phases [51]. The diffraction data collected on both as-sintered Fe80 and Fe50 samples, indicate that, in addition to the non-negligible variation in the amount of extra phases in sample Fe80 upon densification (see Table 1), the main change caused by SPS consists of a significant reduction of the lattice parameter as a consequence of the treatment ($\frac{\Delta a}{a} = -1.5 \times 10^{-4}$ for Fe80 and $\frac{\Delta a}{a} = -2.2 \times 10^{-4}$ for Fe50, where $a$ is the lattice parameter), as can be derived from Figure 3. This result is not obvious. While in fact the diffraction patterns collected under the pressure account for a reduction of the cell volume which depends on the bulk modulus of the material [65], the same does not necessarily apply when the acquisition takes place after the pressure relief. An increase in the lattice parameter versus the applied pressure was for instance reported for $(Ba,DD,Yb)_y(Fe_{1-x}Ni_x)_4Sb_{12}$ [42], $Pr_y(Fe_{1-x}Ni/Co_x)_4Sb_{12}$ and $Ba_yCo_4Sb_{12}$ [33]. Considering that, according to the results of the Rietveld refinements, the filling fraction does not change upon densification, neither does the global composition of the skutterudite (see Table 1), and the observed decrease in the lattice parameter can be ascribed to a memory effect of the material. However, the analysis of the peak width before and after SPS, showing no significant peak broadening, allows the exclusion of any sign of the grain size reduction, at least as far as the technique can reveal. Therefore, relying on the described results, the densification results were mainly responsible for a porosity reduction and for an enhancement of grain connection, which should favor electrical conductivity. No reduction of thermal conductivity due to scattering of phonons on the defects is expected to occur as a consequence of densification. On the contrary, an increase in the lattice thermal conductivity cannot be excluded as a consequence of the reduction in the amount of pores, which constitute an alternative way of phonon scattering.

### 3.4. Thermoelectric Properties

The behavior of the Seebeck coefficient as a function of temperature is reported in Figure 8 for both samples. It can be immediately recognized that the sign of $S$ confirms the $p$- and $n$- nature of Fe80 and Fe50, respectively. Moreover, $|S|$ of the SPS and of the bulk sample are quite close to each other for both compositions as generally expected from samples only differing by their microstructure. The slight enhancement of $|S|$ for SPS with respect to the bulk samples has most probably been attributed to the change in the type and the amount of secondary phases occurring during densification.

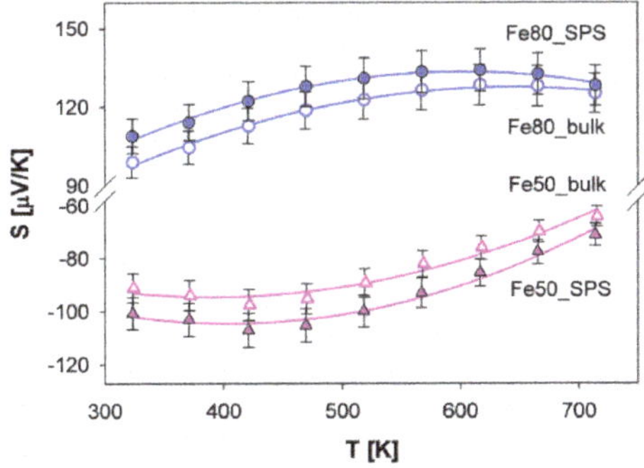


**Figure 8.** The trend of the Seebeck coefficient as a function of temperature for samples Fe80 and Fe50. The curves fitting the experimental points are second order polynomial functions.

The role of the filler on $S$ can be discussed considering that data obtained in this work are essentially comparable to the ones derived from similar systems, such as $Sm_y(Fe_xNi_{1-x})_4Sb_{12}$ [50] and $(Ba,Sr,DD,Yb)_y(Fe_{1-x}Ni_x)_4Sb_{12}$ [42]. Moreover, the room temperature data are in good agreement with the results derived from $(Ce,Yb)_y(Fe_xNi_{1-x})_4Sb_{12}$, $(Ce,Yb)_y(Fe_xCo_{1-x})_4Sb_{12}$, $Ce_y(Fe_xCo_{1-x})_4Sb_{12}$, $Ce_y(Fe_xNi_{1-x})_4Sb_{12}$, $Yb_y(Fe_xCo_{1-x})_4Sb_{12}$ and $Yb_y(Fe_xNi_{1-x})_4Sb_{12}$ systems described in [30], thus suggesting that the Seebeck coefficient is ruled by the charge carriers concentration, rather than by the identity of the filler.

A general feature of semiconductors consists in the occurrence of two different regions in the trend of the absolute value of the Seebeck coefficient versus the temperature, namely an increasing and a decreasing trend at low and high temperatures, respectively, separated by a maximum [42]. This evidence, which is generally accompanied by a similar behavior in the electrical resistivity, can be revealed if the measurement of $S$ extends over a sufficiently broad temperature range, as is the case for our samples. The identification of the maximum value of $S$ ($S_{max}$), as well as of the corresponding temperature ($T_{max}$), is of primary importance, since it allows an estimation of the value of the energy gap ($E_g$), according to the following Equation, developed by Goldsmid and Sharp [66]:

$$S_{max} = \frac{E_g}{2eT_{max}} \qquad (4)$$

being $e$ the electron charge. The data of $T_{max}$, $S_{max}$ and $E_g$ for each sample are collected in Table 3; $T_{max}$ and $S_{max}$ were calculated by fitting experimental data to a second order polynomial function.

**Table 3.** The maximum value of the Seebeck coefficient ($S_{max}$), the temperature corresponding to $S_{max}$ ($T_{max}$), and the value of the energy gap ($E_g$) estimated through Equation (4) for samples Fe80 and Fe50.

| Sample | $S_{max}$ [µV/K] | $T_{max}$ [K] | $E_g$ [meV] |
|---|---|---|---|
| Fe80_bulk | 128 | 642 | 164 |
| Fe80_SPS | 133 | 600 | 160 |
| Fe50_bulk | −97 | 409 | 79 |
| Fe50_SPS | −106 | 413 | 88 |

The trend of the electrical resistivity of samples Fe80 and Fe50 is reported in Figure 9. As a general remark, it can be observed that both compositions exhibited a weak dependence on temperature. Further, despite the weak temperature dependence and the comparatively quite large error bar, a difference can be at a first glance noticed between the two analyzed samples. However, Fe50 shows a semiconducting behavior over the whole temperature range (323 K–723 K), and in Fe80, a plateau-like maximum occurs at a temperature slightly lower than 600 K, meaning that the sample behaves like a bad metal below this temperature, and like a semiconductor above. This is a phenomenon observed at comparable temperatures even in other Fe/Ni-based p-type skutterudites [36,42], which can be explained considering that at lower temperatures the conduction electrons are scattered into unoccupied states. When the conduction band becomes fully occupied, the temperature has to be further increased in order to promote charge carriers across the energy gap, which results in the observed semiconducting behavior at higher temperatures. However, a closer inspection of Figure 9b reveals that in Fe50, the resistivity trend versus the temperature presents two different decreasing rates, being the slope steeper at T< ~400 K than above. This evidence suggests an analogy with the behavior of other *n*-type (Fe/Ni)-based skutterudites, such as (Ba,Sr,DD,Yb)$_y$(Fe$_{1-x}$Ni$_x$)$_4$Sb$_{12}$ (x = 2.1 and 2.2) [42] and DD$_{0.08}$Fe$_2$Ni$_2$Sb$_{12}$ [38], which exhibit a similar slope change between 450 K and 550 K, and a maximum in the 300 K–400 K range. A change in the conduction regime can be thus inferred even for these compositions, not differently from what happens for the corresponding p-type compounds. The temperature at which the maximum in resistivity occurs derives from a subtle interplay among different factors dealing with the density of states at the Fermi level at each temperature, and with the width of the energy gap. The latter contribution can be estimated exploiting the aforementioned dependence of the energy gap ($E_g$) on the maximum value of the Seebeck coefficient versus the temperature ($S_{max}$), i.e., through Equation (4). The calculated values of $E_g$ are 160 meV and 88 meV for samples Fe80_SPS and Fe50_SPS, respectively, as reported in Table 3. The rather smaller value of the calculated $E_g$ for the n-type compound justifies the possible existence of a maximum in the resistivity behavior at a significantly lower temperature than the *p*-type one.

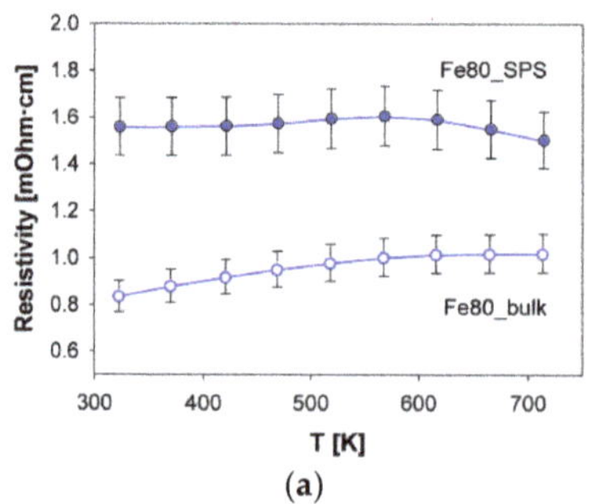


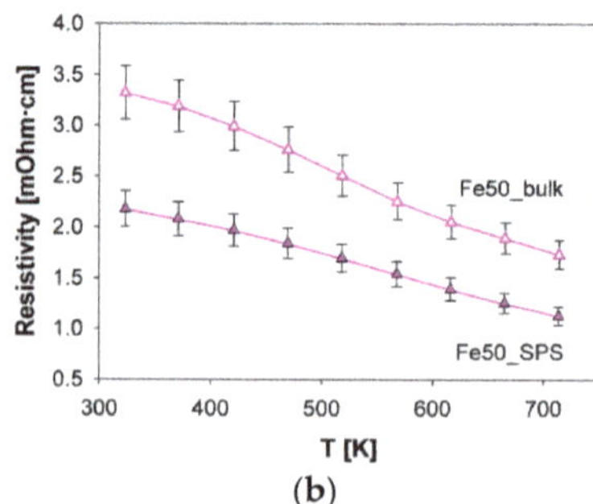


**Figure 9.** The trend of the electrical resistivity as a function of temperature for sample (**a**) Fe80 and (**b**) Fe50. The curves fitting experimental points are splines.

Further, regarding electrical resistivity, it is also worth mentioning that, while in sample Fe50 a higher value occurs for the bulk sample than for the SPS one, the opposite happens for Fe80. This evidence, also noticed for the $Sm_y(Fe_{1-x}Ni_x)_4Sb_{12}$ system [51], is most probably due to the competition between two opposing effects taking place as a consequence of densification, namely an improvement of grain connection, and an increase in the defects amount. While the former promotes a resistivity decrease, the opposite happens for the latter. The behavior observed for Fe80, namely lower resistivity values for the bulk sample, can be ascribed to a couple of different factors. First of all, a possible reason is the relatively scarce density degree reached (88%, significantly lower than 97% obtained for Fe50_SPS, see Table 2), which may favor the predominance of the defect creation over the improvement of the grain connection. Nonetheless, a positive effect of porosity on $ZT$ of $Co_{1-x}Ni_xSb_3$ was already observed by He et al. [67]. Secondarily, since the Fe80 sample contains a non negligible amount of a secondary phase, the electrical resistivity of this sample could be also influenced by the properties of the dispersed secondary phase. Prior to SPS, the sample Fe80_bulk contains a certain amount of antimony (11 wt %), which possesses a relatively low value of electrical resistivity ($\rho_{295.5K} = 34.9\ \mu\Omega$ cm [68]). SPS caused the removal of the majority of antimony, and $(Fe,Ni)Sb_2$, namely a semiconductor with a higher value of electrical resistivity compared to Sb [69] emerged as the dominant secondary phase. The reduced amount of metallic Sb together with the increased amount of semiconducting $(Fe,Ni)Sb_2$ is a further possible reason why the electrical resistivity is higher in Fe80_SPS than in Fe80_bulk.

Combining electrical conductivity and the Seebeck coefficient allows the calculation of the power factor ($PF$) at each temperature considered. In Figure 10, the $PF$ values of the samples Fe80 and Fe50 are reported. For both compositions, the temperature at which the maximum power factor occurs is slightly higher than the one corresponding to the maximum value of the Seebeck coefficient. The obtained trends of $PF$ versus the temperature can be compared to the data deriving from similar Sb-based filled skutterudites, such as $(Ba,Sr,DD,Yb)_y(Fe_{1-x}Ni_x)_4Sb_{12}$ [42]. It can be observed that values reported for the aforementioned system both at $x = 0.50$ and 0.80 are higher than the ones claimed in this work.

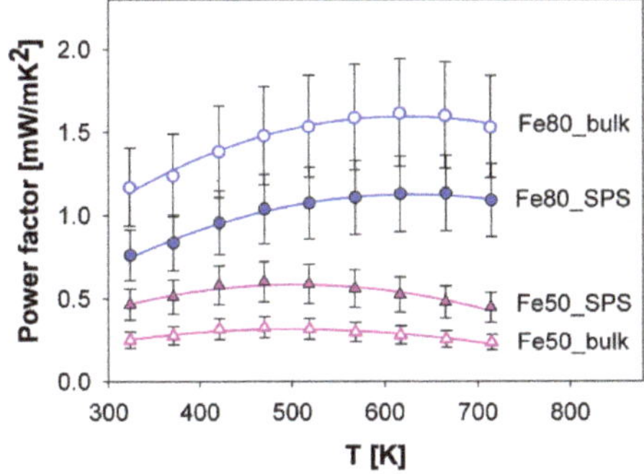


**Figure 10.** The trend of the power factor as a function of the temperature for samples Fe80 and Fe50. The curves fitting the experimental points are second order polynomial functions.

This evidence can be most probably be ascribed to the lower values of electrical resistivity of the multi-filled compounds, rather than to the Seebeck coefficient. As previously described, in fact, $S$ values of the systems containing even different fillers are quite close to each other, pointing at a dependence of the Seebeck coefficient mainly on electronic issues. This interpretation is confirmed by the analysis of the data collected at 350 K on $(Sm,Gd)_{0.15}(Fe_{0.5}Ni_{0.5})_4Sb_{12}$ (this work) and $Yb_y(Fe_{0.5}Ni_{0.5})_4Sb_{12}$ [43]. At this temperature, the former system is characterized by a higher power factor when compared to each composition of the latter. Again, the electrical resistivity seems to be responsible for this dissimilarity, being significantly lower in $(Sm,Gd)_{0.15}(Fe_{0.5}Ni_{0.5})_4Sb_{12}$ than in $Yb_y(Fe_{0.5}Ni_{0.5})_4Sb_{12}$. It can be thus concluded that the improvement of the power factor is mainly related to the improvement of electrical conductivity, which can be accomplished by optimizing the material processing.

The trends of overall ($\lambda_{ov}$), and phonon and bipolar ($\lambda_{ph} + \lambda_{bp}$) thermal conductivity of Fe80 and Fe50 are reported in Figure 11. While ($\lambda_{ph} + \lambda_{bp}$) of Fe50 does not significantly change with densification, it increases for Fe80. The latter behavior is generally expected, since pores can offer a further way for phonon scattering.

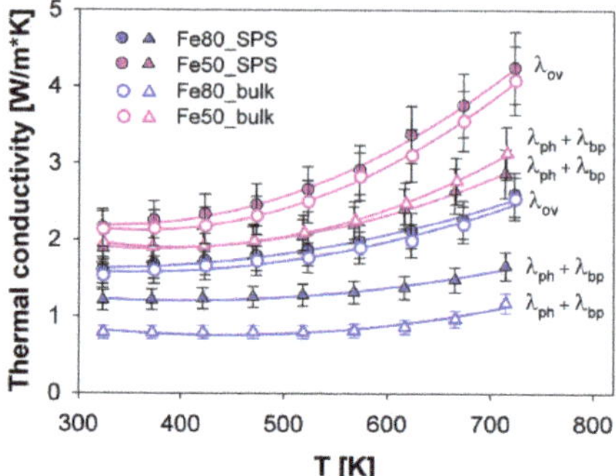


**Figure 11.** The trend of overall ($\lambda_{ov}$), and lattice and bipolar ($\lambda_{ph} + \lambda_{bp}$) thermal conductivity of the samples Fe80_SPS and Fe50_SPS as a function of the temperature.

It can be observed that $\lambda_{ov}$ increases with increasing temperature for both samples, but more steeply for Fe50_SPS, with 1.6 W/m·K and 4.2 W/m·K the minimum and maximum value in the 323 K–723 K temperature range. As already revealed for $DD_y(Fe_xNi_{1-x})_4Sb_{12}$ [38], even in the $(Sm,Gd)_y(Fe_xNi_{1-x})_4Sb_{12}$ system, thermal conductivity is lower at $x = 0.8$ than at x = 0.5. This result can be understood considering that the thermal conductivity reduction arises from the two-fold effect of a) the rattling motion of the filler, which is responsible for the $\lambda$ reduction at low temperatures, and b) the disorder brought about by the contemporary presence of two different atoms at the 8c position (Fe and Ni in the present study) [70]. By relying on the data reported in [70], this combined effect causes the minimum lattice thermal conductivity to be located at $x \sim 1.5$ in $Ce_yFe_{4-x}Co_xSb_{12}$, which corresponds to $x = 0.80$ in $(Sm,Gd)_y(Fe_xNi_{1-x})_4Sb_{12}$, if Fe is taken as a bivalent and Ni as a tetravalent ion, as suggested in [18].

The effect of the sole filler on thermal conductivity can be studied by comparing antimonides having the same (Fe/Ni) amount ratio. For instance, the comparison between $\lambda_{ov}$ of $(Sm,Gd)_y(Fe_xNi_{1-x})_4Sb_{12}$ and $Sm_y(Fe_xNi_{1-x})_4Sb_{12}$ [50] with x = 0.8, points at a lower value for the former system, as a consequence of double filling. An analogous comparative approach applied to $\lambda_{ov}$ of $(Sm,Gd)_y(Fe_xNi_{1-x})_4Sb_{12}$ with x = 0.5 and $Yb_y(Fe_{0.5}Ni_{0.5})_4Sb_{12}$ [43] reveals similar values in the two systems, at least in the 350 K–650 K temperature range. This evidence, which at a first glance could seem contradictory, finds an explanation when the valence state of Yb is taken into account. According to X-ray absorption spectroscopy measurements performed on $(Ce-Yb)_yFe_{4-x}Co_xSb_{12}$ and $(Ce-Yb)_yFe_{4-x}Ni_xSb_{12}$ [30], Yb occurs in a +2/+3 mixed valence state, evolving toward +3 with decreasing the filling degree. This intermediate valence state leads a Yb-containing skutterudite to mimic a double-filled system, as a consequence of the different size of $Yb^{2+}$ and $Yb^{3+}$, which can justify the observed very similar values of thermal conductivity of the Yb- and the (Sm,Gd)-filled skutterudite.

Assuming the validity of the Wiedemann-Franz law:

$$\lambda_{el} = \frac{L_0 T}{\rho} \tag{5}$$

where $L_0$ is the Lorenz number (calculated using the Seebeck coefficient, according to [71]), it is possible to give an estimation of the sum of lattice and bipolar thermal conductivity ($\lambda_{ph} + \lambda_{bp}$) by subtracting $\lambda_{el}$ from $\lambda_{ov}$. The values reported in Figure 11 were obtained this way. It is noteworthy that $(\lambda_{ph} + \lambda_{bp})$ increases with increasing temperature similarly to $\lambda_{ov}$, but the increasing rate is much weaker, particularly in the high temperature region. This result suggests that with increasing temperature $\lambda_{el}$ becomes the main contribution to thermal conductivity. If $\lambda_{ph}$ of the $(Ba,DD,Yb)_y(Fe_{1-x}Ni_x)_4Sb_{12}$

system with x = 0.78 [42] is considered, it can be noticed that the values are very close to the ones of Fe80_bulk, in what can be deemed as a very promising result, since it is reached making use of just two fillers.

A closer inspection of the $(\lambda_{ph} + \lambda_{bp})$ trend versus the temperature allows a deeper insight into the role of the bipolar contribution to thermal conductivity of the present system. The cited factor is determined by the electron-hole pairs, which together with electrons and phonons are responsible for the heat transfer in semiconductors. The electron-hole coupling occurs at sufficiently high temperatures depending on the band gap, so that, as aforementioned, $\lambda_{bp}$ becomes relevant only above a certain threshold, which is strictly bound to the system under examination. A useful way to reveal the signature of the electron-hole coupling is to plot $(\lambda_{ov} - \lambda_{el})$ as a function of $T^{-1}$. At low temperatures, a decreasing linear trend with increasing temperature is generally observed due to Umklapp scattering, and the temperature where data start deviating from linearity can be taken as the threshold temperature above which the bipolar contribution becomes significant. The data deriving from Yb-filled $Co_4Sb_{12}$ follow this model [72].

Figure 12a reports the described plot for samples Fe80_SPS and Fe50_SPS. It can be seen that $(\lambda_{ph} + \lambda_{bp})$ of both samples tends to increase with increasing temperature, instead of decreasing with the usual $T^{-1}$ temperature dependence, as previously described. Further, above ~570 K for Fe80_SPS and ~470 K for Fe50_SPS, the values start growing more steeply with increasing temperature, in good agreement with the larger $E_g$ calculated for the former composition (160 meV) with respect to the latter (88 meV). However, the increasing trend along the whole temperature range suggests that the bipolar contribution near room temperature is already significant and able to mask the contribution of Umklapp scattering. This is particularly relevant for the Fe50_SPS sample, since the extracted thermal band gap is only 88 meV, which is just a few kT at room temperature. In order to quantify the contribution of both processes, a least square fit of the experimental data was performed using the relation for bipolar thermal conductivity

$$\lambda_{bp} = cT^n exp\left(-\frac{E_g}{2kT}\right) \tag{6}$$

where $c$ and $n$ are adjustable parameters [4], together with the aforementioned $T^{-1}$ dependency for the high temperature Umklapp scattering. A reasonably good agreement was observed for the Fe50_SPS sample, as shown in Figure 12b. As expected, the contribution of bipolar conductivity amounts to approximately 10% near room temperature and increases significantly at higher temperature. The same approach does not work satisfactorily on Fe80_SPS, presumably due to the presence of secondary phases, that are not accounted for by the described model.

The trend of $ZT$ as a function of the temperature for both Fe80 and Fe50 is shown in Figure 13. First, it can be noticed that Fe80 is characterized by higher $ZT$ values than Fe50 at each temperature, as a consequence of the higher power factor and the lower total thermal conductivity, as previously described. Moreover, the maximum $ZT$ occurs at ~615 K for Fe80, and at ~500 K for Fe50. Interestingly, Fe80_bulk shows higher values of $ZT$ than Fe80_SPS, due to the lower thermal conductivity and to the higher values of the power factor, which are in turn caused by the previously described lower resistivity of the porous sample with respect to the dense one. The comparison between $ZT$ data of Fe80 and the ones of the corresponding Sm-filled sample [50] indicates that the single-filled system presents the maximum at a slightly higher temperature (~615 K), and a lower value of the figure of merit, as a consequence of the higher thermal conductivity. Analogously, a slightly lower value of the figure of merit is shown by $Yb_y(Fe_{0.5}Ni_{0.5})_4Sb_{12}$ at each y value [43] with respect to Fe50, as could be expected from the aforementioned very similar values of thermal conductivity for the two systems. Finally, the higher value of the maximum $ZT$ occurring for the $(Ba,DD,Yb)_y(Fe_{1-x}Ni_x)_4Sb_{12}$ system with $x = 0.78$ [42], can be considered as a direct consequence of the previously discussed higher power factor value.

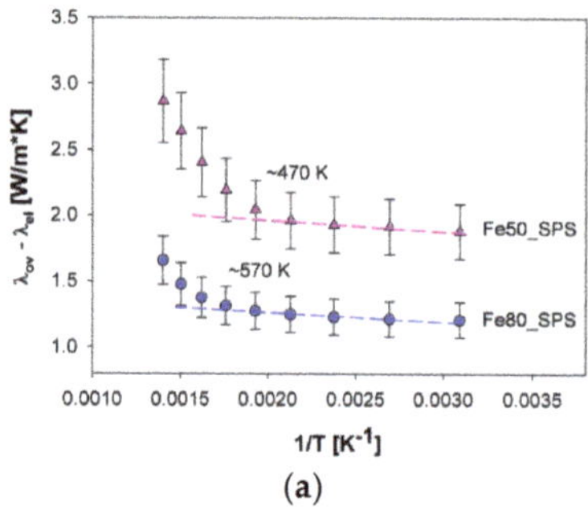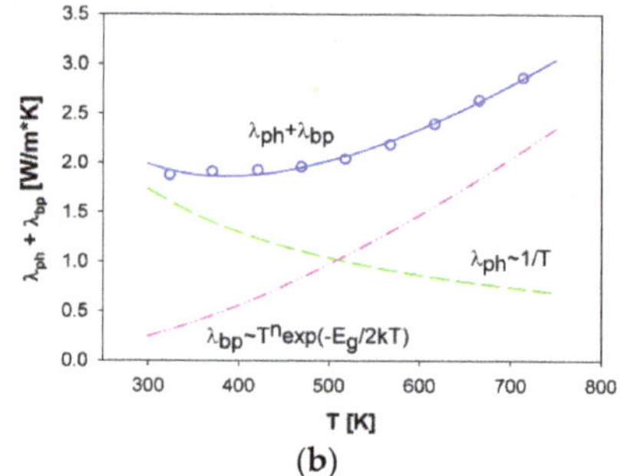

**Figure 12.** (**a**) The trend of ($\lambda_{ov} - \lambda_{el}$) vs. 1/T for samples Fe80_SPS and Fe50_SPS. The dashed lines are a guide for the eye. The temperatures reported within the diagram roughly indicate the threshold above which the bipolar contribution starts growing steeply; (**b**) fit of ($\lambda_{ph} + \lambda_{bp}$) data (continuous line) performed according to Equation 6) (dashed-dotted line) and to Umklapp scattering (dashed line).

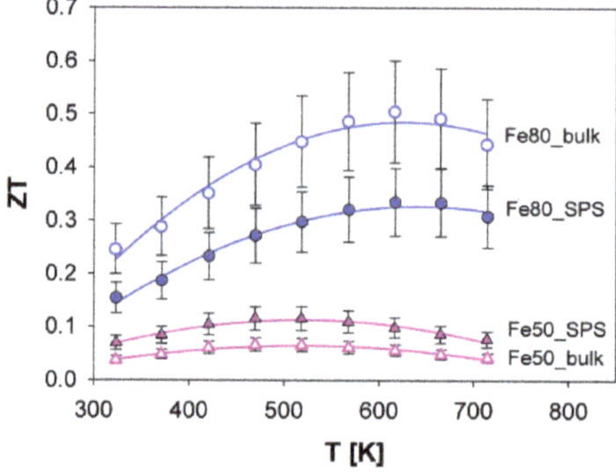

**Figure 13.** The trend of $ZT$ as a function of the temperature for the samples Fe80 and Fe50. The curves fitting experimental points are second order polynomial functions.

The closeness between the studied system and the described multi-filled (Fe,Ni)-based skutterudites in terms of thermoelectric performance, and in particular of thermal conductivity, allows the conclusion that the selection of the (Sm,Gd) mixture as a filler is a good compromise between the limited $Sm^{3+}/Gd^{3+}$ mass difference and the relatively low filling fraction.

## 4. Conclusions

The structural and thermoelectric properties of both *p*- and *n*- compositions belonging to the skutterudite system $(Sm,Gd)_y(Fe_xNi_{1-x})_4Sb_{12}$ were studied with the aim of evaluating the effect of double filling by Sm and Gd on composition, crystallographic parameters and transport properties of the material.

The results of the structural investigation suggest that the employment of the (Sm,Gd) mixture as a filler causes the filling fraction to be always lower when using Ce or Sm. Moreover, with increasing the Fe amount, the Sm/(Sm + Gd) ratio increases too, suggesting that Sm is preferred when the cage size grows. Both evidences point at a role of the filler size in ruling the filling fraction, in addition to the filler oxidation state.

Despite the $Sm^{3+}/Gd^{3+}$ small mass difference, the outcome of the thermoelectric study shows that the contemporary presence of these ions within the void located in the 2*a* site significantly reduces at each temperature considered by the total thermal conductivity of both *p*- and *n*-compositions below the values of the only Sm-filled compound. Moreover, both the phonon thermal conductivity and the overall thermoelectric performance are comparable to those of several multi-filled (Fe,Ni)-based skutterudite antimonides described in the literature. The choice of the (Sm,Gd) mixture as a filler thus results to be a good compromise between the limitations imposed by the reduced $Gd^{3+}$ size and the $Sm^{3+}/Gd^{3+}$ mass closeness.

**Author Contributions:** Conceptualization, C.A., R.C. and P.M.; methodology, C.A., F.F. and T.M.; formal analysis, C.A. and F.F.; investigation, R.C., R.S. and F.F.; resources, R.C.; writing—original draft preparation, C.A.; writing—review and editing, all; visualization, C.A.; funding acquisition, T.M.

**Funding:** This research was funded by CREST, grant number JPMJCR15Q6, and JSPS, grant number KAKENHI JP17H02749.

**Acknowledgments:** Cristina Artini and Paolo Mele gratefully thank Shrikant Saini, Atsunori Kamegawa and Toru Kimura of Research Center for Environmentally Friendly Materials Engineering, Muroran Institute of Technology (Muroran, Japan), for providing ball milling facility and giving assistance in the ball milling experiment; Prof. Tsunehiro Takeuchi and Seongho Choi of Toyota Technological Institute (Nagoya, Japan) for providing the samples densification with SPS.

**Conflicts of Interest:** The authors declare no conflict of interest.

## References


1. Mori, T.; Priya, S. Materials for energy harvesting: At the forefront of a new wave. *MRS Bull.* **2018**, *43*, 176–180. [CrossRef]

2. Petsagkourakis, I.; Tybrandt, K.; Crispin, X.; Ohkubo, I.; Satoh, N.; Mori, T. Thermoelectric materials and applications for energy harvesting power generation. *Sci. Tech. Adv. Mater.* **2018**, *19*, 836–862. [CrossRef] [PubMed]

3. Wang, S.; Yang, J.; Toll, T.; Yang, J.; Zhang, W.; Tang, X. Conductivity-limiting bipolar thermal conductivity in semiconductors. *Sci. Rep.* **2015**, *5*, 10136. [CrossRef] [PubMed]

4. Ceyda Yelgel, Ö.; Srivastava, G.P. Thermoelectric properties of $n$-type $Bi_2(Te_{0.85}Se_{0.15})_3$ single crystals doped with CuBr and $SbI_3$. *Phys. Rev. B* **2012**, *85*, 125207. [CrossRef]

5. You, Y.; Su, X.; Liu, W.; Yan, Y.; Fu, J.; Cheng, X.; Zhang, C.; Tang, X. Structure and thermoelectric property of Te doped paracostibite $CoSb_{1-x}Te_xS$ compounds. *J. Solid State Chem.* **2018**, *262*, 1–7. [CrossRef]

6. Carlini, R.; Artini, C.; Borzone, G.; Masini, R.; Zanicchi, G.; Costa, G.A. Synthesis and characterization of the compound CoSbS. *J. Therm. Anal. Calorim.* **2011**, *103*, 23–27. [CrossRef]

7. Basit, A.; Yang, J.; Jiang, Q.; Zhou, Z.; Xin, J.; Li, X.; Li, S. Effect of Sn doping on thermoelectric properties of $p$-type manganese telluride. *J. Alloy Compd.* **2019**, *777*, 968–973. [CrossRef]

8. Liu, Z.; Mao, J.; Liu, T.H.; Chen, G.; Ren, Z. Nano-microstructural control of phonon engineering for thermoelectric energy harvesting. *MRS Bull.* **2018**, *43*, 181–186. [CrossRef]

9. Dura, O.J.; Andujar, R.; Falmbigl, M.; Rogl, P.; Lòpez de la Torre, M.A.; Bauer, E. The effect of nanostructure on the thermoelectric figure-of-merit of $La_{0.875}Sr_{0.125}CoO_3$. *J. Alloy Compd.* **2017**, *711*, 381–386. [CrossRef]

10. Xu, W.; Liu, Y.; Chen, B.; Liu, D.B.; Lin, Y.H.; Marcelli, A. Nano-inclusions: A novel approach to tune the thermal conductivity of $In_2O_3$. *Phys. Chem. Chem. Phys.* **2013**, *15*, 17595–17600. [CrossRef]

11. Srinivasan, B.; Fontaine, B.; Gucci, F.; Dorcet, V.; Graves Saunders, T.; Yu, M.; Cheviré, F.; Boussard-Pledel, C.; Halet, J.F.; Gautier, R.; et al. Effect of the processing route on the thermoelectric performance of nanostructured $CuPb_{18}SbTe_{20}$. *Inorg. Chem.* **2018**, *57*, 12976–12986. [CrossRef]

12. Liu, W.; Shi, X.; Moshwan, R.; Hong, M.; Yang, L.; Chen, Z.-G.; Zou, J. Enhancing thermoelectric performance of $(Cu_{1-x}Ag_x)_2Se$ via CuAgSe secondary phase and porous design. *Sust. Mater. Tech.* **2018**, *17*, e00076. [CrossRef]

13. Birkel, C.S.; Douglas, J.E.; Lettiere, B.R.; Seward, G.; Verma, N.; Zhang, Y.; Pollock, T.M.; Seshadri, R.; Stucky, G.D. Improving the thermoelectric properties of half-Heusler TiNiSn through inclusion of a second full-Heusler phase: microwave preparation and spark plasma sintering of $TiNi_{1+x}Sn$. *Phys. Chem. Chem. Phys.* **2013**, *15*, 6990–6997. [CrossRef]

14. Ferluccio, D.A.; Smith, R.I.; Buckman, J.; Bos, J.-W.G. Impact of Nb vacancies and p-type doping of the NbCoSn-NbCoSb half-Heusler thermoelectric. *Phys. Chem. Chem. Phys.* **2018**, *20*, 3979–3987. [CrossRef] [PubMed]

15. Khan, A.U.; Kobayashi, K.; Tang, D.-M.; Yamauchi, Y.; Hasegawa, K.; Mitome, M.; Xue, Y.; Jiang, B.; Tsuchiya, K.; Golberg, D.; et al. Nano-micro-porous skutterudites with 100% enhancement in ZT for high performance thermoelectricity. *Nano Energy* **2017**, *31*, 152–159. [CrossRef]

16. Mori, T. Novel principles and nanostructuring methods for enhanced thermoelectrics. *Small* **2017**, *13*, 1702013. [CrossRef] [PubMed]

17. Slack, G.A. New Materials and Performance Limits for Thermoelectric Cooling. In *CRC Handbook of Thermoelectrics*, 1st ed.; Rowe, D.M., Ed.; Taylor and Francis: Boca Raton, FL, USA, 1995; pp. 407–439.

18. Uher, C. Skutterudite-Based Thermoelectrics. In *Thermoelectrics Handbook—Macro to Nano*, 1st ed.; Rowe, D.M., Ed.; Taylor and Francis: Boca Raton, FL, USA, 2005; Chapter 34; pp. 1–17.

19. Dolyniuk, J.A.; Owens-Baird, B.; Wang, J.; Zaikina, J.V.; Kovnir, K. Clathrate thermoelectrics. *Mater. Sci. Eng. R* **2016**, *108*, 1–46. [CrossRef]

20. Abramchuk, N.S.; Carrillo-Cabrera, W.; Veremchuk, I.; Oeschler, N.; Olenev, A.V.; Prots, Y.; Burkhardt, U.; Dikarev, E.V.; Grin, J.; Shevelkov, A.V. Homo- and Heterovalent Substitutions in the New Clathrates I $Si_{30}P_{16}Te_{8-x}Se_x$ and $Si_{30+x}P_{16-x}Te_{8-x}Br_x$: Synthesis, Crystal Structure, and Thermoelectric Properties. *Inorg. Chem.* **2012**, *51*, 11396–11405. [CrossRef]

21. Chen, S.; Ren, Z. Recent progress of half-Heusler for moderate temperature thermoelectric applications. *Mater. Today* **2013**, *16*, 387–395. [CrossRef]

22. Sales, B.C. Filled skutterudites. In *Handbook on the Physics and Chemistry of Rare Earths*, 1st ed.; Gschneidner, K.A., Jr., Bünzli, J.-C.G., Pecharsky, V.K., Eds.; North Holland: Amsterdam, The Netherlands, 2003; Volume 33, pp. 1–34.

23. Chakoumakos, B.C.; Sales, B.C. Skutterudites: Their structural response to filling. *J. Alloy Compd.* **2006**, *407*, 87–93. [CrossRef]

24. Uher, C. In search of efficient n-type skutterudite thermoelectrics. In Proceedings of the 21st Conference on Thermoelectrics, Long Beach, CA, USA, 29 August 2002; IEEE: Piscataway, NJ, USA, 2002; pp. 35–41.

25. Keppens, V.; Mandrus, D.; Sales, B.C.; Chakoumakos, B.C.; Dai, P.; Coldea, R.; Maple, M.B.; Gajewski, D.A.; Freeman, E.J.; Bennington, S. Localized vibrational modes in metallic solids. *Nature* **1998**, *395*, 876–878. [CrossRef]

26. Sootsman, J.R.; Chung, D.Y.; Kanatzidis, M. New and old concepts in thermoelectric materials. *Angew. Chem. Int. Edit.* **2009**, *48*, 8616–8639. [CrossRef]

27. Tang, Y.; Chen, S.; Jeffrey-Snyder, G. Temperature dependent solubility of Yb in Yb-CoSb$_3$ skutterudite and its effect on preparation, optimization and lifetime of thermoelectrics. *J. Materiomics.* **2015**, *1*, 75–84. [CrossRef]

28. Liu, R.; Cho, J.Y.; Yang, J.; Zhang, W.; Chen, L. Thermoelectric transport properties of $R_yFe_3NiSb_{12}$ (R = Ba, Nd and Yb). *J. Mater. Sci. Technol.* **2014**, *30*, 1134–1140. [CrossRef]

29. Chapon, L.; Ravot, D.; Tedenac, J.C. Nickel-substituted skutterudites: synthesis, structural and electrical properties. *J. Alloy Compd.* **1999**, *282*, 58–63. [CrossRef]

30. Bérardan, D.; Alleno, E.; Godart, C.; Rouleau, O. Rodriguez-Carvajal, Preparation and chemical properties of the skutterudites $(Ce-Yb)_yFe_{4-x}(Co/Ni)_xSb_{12}$. *Mater. Res. Bull.* **2005**, *40*, 537–551. [CrossRef]

31. Bourgouin, B.; Bérardan, D.; Alleno, E.; Godart, C.; Rouleau, O.; Leroy, E. Preparation and thermopower of new mischmetal-based partially filled skutterudites $Mm_yFe_{4-x}(Co/Ni)_xSb_{12}$. *J. Alloy Compd.* **2005**, *399*, 47–51. [CrossRef]

32. Rogl, G.; Grytsiv, A.; Falmbigl, M.; Bauer, E.; Rogl, P.; Zehetbauer, M.; Gelbstein, Y. Thermoelectric properties of p-type didymium (DD) based skutterudites $DD_y(Fe_{1-x}Ni_x)_4Sb_{12}$ (0.13 ≤ x ≤ 0.25, 0.46 ≤ y ≤ 0.68). *J. Alloy Compd.* **2012**, *537*, 242–249. [CrossRef]

33. Zhang, L.; Grytsiv, A.; Bonarski, B.; Kerber, M.; Setman, D.; Schafler, E.; Rogl, P.; Bauer, E.; Hilscher, G.; Zehetbauer, M. Impact of high pressure torsion on the microstructure and physical properties of $Pr_{0.67}Fe_3CoSb_{12}$, $Pr_{0.71}Fe_{3.5}Ni_{0.5}Sb_{12}$, and $Ba_{0.06}Co_4Sb_{12}$. *J. Alloy Compd.* **2010**, *494*, 78–83. [CrossRef]

34. Artini, C.; Zanicchi, G.; Costa, G.A.; Carnasciali, M.M.; Fanciulli, C.; Carlini, R. Correlations between structural and electronic properties in the filled skutterudite $Sm_y(Fe_xNi_{1-x})_4Sb_{12}$. *Inorg. Chem.* **2016**, *55*, 2574–2583. [CrossRef]

35. Artini, C.; Fanciulli, C.; Zanicchi, G.; Costa, G.A.; Carlini, R. Thermal expansion and high temperature structural features of the filled skutterudite $Sm_\beta(Fe_\alpha Ni_{1-\alpha})_4Sb_{12}$. *Intermetallics* **2017**, *87*, 31–37. [CrossRef]

36. Rogl, G.; Grytsiv, A.; Heinrich, P.; Bauer, E.; Kumar, P.; Peranio, N.; Eibl, O.; Horky, J.; Zehetbauer, M.; Rogl, P. New bilk p-type skutterudites $DD_{0.7}Fe_{2.7}Co_{1.3}Sb_{12-x}X_x$ (X = Ge, Sn) reaching ZT > 1.3. *Acta Mater.* **2015**, *91*, 227–238. [CrossRef]

37. Duan, F.; Zhang, L.; Dong, J.; Sakamoto, J.; Xu, B.; Li, X.; Tian, Y. Thermoelectric properties of Sn substituted p-type Nd filled skutterudites. *J. Alloy Compd.* **2015**, *639*, 68–73. [CrossRef]

38. Rogl, G.; Grytsiv, A.; Bauer, E.; Rogl, P.; Zehetbauer, M. Thermoelectric properties of novel skutterudites with didymium: $DD_y(Fe_{1-x}Co_x)_4Sb_{12}$ and $DD_y(Fe_{1-x}Ni_x)_4Sb_{12}$. *Intermetallics* **2010**, *18*, 57–64. [CrossRef]

39. Kitagawa, H.; Hasaka, M.; Morimura, T.; Nakashima, H.; Kondo, S. Skutterudite structure and thermoelectric property in $Ce_fFe_{8-x}Co_xSb_{24}$ (f = 0-2, x = 0-8). *Mater. Res. Bull.* **2000**, *35*, 185–192. [CrossRef]

40. Kong, L.; Jia, X.; Zhang, Y.; Sun, B.; Liu, B.; Liu, H.; Wang, C.; Liu, B.; Chen, J.; Ma, H. N-type $Ba_{0.3}Ni_{0.15}Co_{3.85}Sb_{12}$ skutterudite: High pressure processing technique and thermoelectric properties. *J. Alloy Compd.* **2018**, *734*, 36–42. [CrossRef]

41. Zhang, L.; Rogl, G.; Grytsiv, A.; Puchegger, S.; Koppensteiner, J.; Spieckermann, F.; Kabelka, H.; Reinecker, M.; Rogl, P.; Schranz, W.; et al. Mechanical properties of filled antimonide skutterudites. *Mater. Sci. Eng. B* **2010**, *170*, 26–31. [CrossRef]

42. Rogl, G.; Grytsiv, A.; Royanian, E.; Heinrich, P.; Bauer, E.; Rogl, P.; Zehetbauer, M.; Puchegger, S.; Reinecker, M.; Schranz, W. New p and n-type skutterudites with ZT>1 and nearly identical thermal expansion and mechanical properties. *Acta Mater.* **2013**, *61*, 4066–4079. [CrossRef]

43. Kaltzoglou, A.; Vaqueiro, P.; Knight, K.S.; Powell, A.V. Synthesis, characterization and physical properties of the skutterudites $Yb_xFe_2Ni_2Sb_{12}$ ($0 \leq x \leq 0.4$). *J. Solid State Chem.* **2012**, *193*, 36–41. [CrossRef]

44. Dahal, T.; Jie, Q.; Lan, Y.; Guo, C.; Ren, Z. Thermoelectric performance of Ni compensated cerium and neodymium double filled p-type skutterudites. *Phys. Chem. Chem. Phys.* **2014**, *16*, 18170–18175. [CrossRef]

45. Ballikaya, S.; Uzar, N.; Yildirim, S.; Salvador, J.R.; Uher, C. High thermoelectric performance of In, Yb, Ce multiple filled $CoSb_3$ based skutterudite compounds. *J. Solid State Chem.* **2012**, *193*, 31–35. [CrossRef]

46. Rogl, G.; Zhang, L.; Rogl, P.; Grytsiv, A.; Falmbigl, M.; Rajs, D.; Kriegisch, M.; Müller, H.; Bauer, E.; Koppensteiner, J.; et al. Thermal expansion of skutterudites. *J. Appl. Phys.* **2010**, *107*, 043507. [CrossRef]

47. Matsubara, M.; Masuoka, Y.; Asahi, R. Effects of doping IIIB elements (Al, Ga, In) on thermoelectric properties of nanostructured n-type filled skutterudite compounds. *J. Alloy Compd.* **2019**, *774*, 731–738. [CrossRef]

48. Karthikeyan, N.; Sivaprasad, G.; Jaiganesh, G.; Anbarasu, V.; Partha Pratim, J.; Sivakumar, K. Thermoelectric properties of Se and Zn/Cd/Sn double substituted $Co_4Sb_{12}$ skutterudite compounds. *Phys. Chem. Chem. Phys.* **2017**, *19*, 28116–28126.

49. Artini, C.; Castellero, A.; Baricco, M.; Buscaglia, M.T.; Carlini, R. Structure, microstructure and microhardness of rapidly solidified $Sm_y(Fe_xNi_{1-x})_4Sb_{12}$ (x = 0.45, 0.50, 0.70, 1) thermoelectric compounds. *Solid State Sci.* **2018**, *79*, 71–78. [CrossRef]

50. Carlini, R.; Khan, A.U.; Ricciardi, R.; Mori, T.; Zanicchi, G. Synthesis, characterization and thermoelectric properties of Sm filled $Fe_{4-x}Ni_xSb_{12}$. *J. Alloy Compd.* **2016**, *655*, 321–326. [CrossRef]

51. Artini, C.; Latronico, G.; Carlini, R.; Saini, S.; Takeuchi, T.; Choi, S.; Baldini, A.; Anselmi-Tamburini, U.; Valenza, F.; Mele, P. Effect of different processing routes on the power factor of the filled skutterudite $Sm_y(Fe_xNi_{1-x})_4Sb_{12}$. *ES Mater. Manuf.* **2019**. [CrossRef]

52. Artini, C.; Parodi, N.; Latronico, G.; Carlini, R. Formation and decomposition process of the filled skutterudite $Sm_y(Fe_xNi_{1-x})_4Sb_{12}$ ($0.40 \leq x \leq 1$) as revealed by differential thermal analysis. *J. Mater. Eng. Perform.* **2018**, *27*, 6259–6265. [CrossRef]

53. Artini, C.; Carlini, R. Influence of composition and thermal treatments on microhardness of the filled skutterudite $Sm_y(Fe_xNi_{1-x})_4Sb_{12}$. *J. Nanosci. Nanotechnol.* **2017**, *17*, 1634–1639. [CrossRef]

54. Carlini, R.; Parodi, N.; Soggia, F.; Latronico, G.; Carnasciali, M.M.; Artini, C. Corrosion behavior of $Sm_y(Fe_xNi_{1-x})_4Sb_{12}$ ($0.40 \leq x \leq 0.80$) in sodium chloride solutions studied by electron microscopy and ICP-AES. *J. Mater. Eng. Perform.* **2018**, *27*, 6266–6273. [CrossRef]

55. Weast, R.C. *Handbook of Chemistry and Physics*, 56th ed.; CRC Press: Cleveland, OH, USA, 1975.

56. Schindelin, J.; Arganda-Carreras, I.; Frise, E.; Kaynig, V.; Longair, M.; Pietzsch, T.; Preibisch, S.; Rueden, C.; Saalfeld, S.; Schmid, B.; et al. Fiji: an open-source platform for biological-image analysis. *Nat. Methods* **2012**, *9*, 676–682. [CrossRef]

57. Schneider, C.A.; Rasband, W.S.; Eliceiri, K.W. NIH Image to ImageJ: 25 years of image analysis. *Nat. Methods* **2012**, *9*, 671–675. [CrossRef]

58. Rodriguez-Carvajal, J. Recent advances in magnetic structure determination by neutron powder diffraction. *Physica B* **1993**, *192*, 55–69. [CrossRef]

59. Alleno, E.; Bérardan, D.; Byl, C.; Candolfi, C.; Daou, R.; Decourt, R.; Guilmeau, E.; Hébert, S.; Hejtmanek, J.; Lenoir, B.; et al. A round robin test of the uncertainty on the measurement of the thermoelectric dimensionless figure of merit of $Co_{0.97}Ni_{0.03}Sb_3$. *Rev. Sci. Instr.* **2015**, *86*, 011301. [CrossRef]

60. Shannon, R.D. Revised effective ionic radii and systematic studies of interatomic distances in halides and chalcogenides. *Acta Crystallogr. A* **1976**, *32*, 751–767. [CrossRef]

61. Morimura, T.; Hasaka, M. Partially filled skutterudite structure in $Ce_fFe_{8-x}Ni_xSb_{24}$. *Scr. Mater.* **2003**, *48*, 495–500.

62. Shi, X.; Kong, H.; Li, C.-P.; Uher, C.; Yang, J.; Salvador, J.R.; Wang, H.; Chen, L.; Zhang, W. Low thermal conductivity and high thermoelectric figure of merit in $n$-type $Ba_xYb_yCo_4Sb_{12}$ double-filled skutterudites. *Appl. Phys. Lett.* **2008**, *92*, 182101. [CrossRef]

63. Rogl, G.; Grytsiv, A.; Bursik, J.; Horky, J.; Anbalagan, R.; Bauer, E.; Mallik, R.C.; Rogl, P.; Zehetbauer, M. Changes in microstructure and physical properties of skutterudites after severe plastic deformation. *Phys. Chem. Chem. Phys.* **2015**, *17*, 3715–3722. [CrossRef]

64. Li, X.; Zhang, Q.; Kang, Y.; Chen, C.; Zhang, L.; Yu, D.; Tian, Y.; Xu, B. High pressure synthesized Ca-filled $CoSb_3$ skutterudites with enhanced thermoelectric properties. *J. Alloy Compd.* **2016**, *677*, 61–65. [CrossRef]

65. Artini, C.; Joseph, B.; Costa, G.A.; Pani, M. Crystallographic properties of the $Ce_{1-x}Lu_xO_{2-x/2}$ system at pressures up to 7 GPa. *Solid State Ionics* **2018**, *320*, 152–158. [CrossRef]

66. Goldsmid, H.J.; Sharp, J.W. Estimation of the thermal band gap of a semiconductor from Seebeck measurements. *J. Electron. Mater.* **1999**, *28*, 869–872. [CrossRef]

67. He, Q.; Hu, S.; Tang, X.; Lan, Y.; Yang, J.; Wang, X.; Ren, Z.; Hao, Q.; Chen, G. The great improvement effect of pores on ZT in $Co_{1-x}Ni_xSb_3$ system. *Appl. Phys. Lett.* **2008**, *93*, 042108. [CrossRef]

68. Horne, R.A. Effect of oxide impurities on the thermoelectric powers and electrical resistivities of bismuth, antimony, tellurium, and bismuth-tellurium alloys. *J. Appl. Phys.* **1959**, *30*, 393. [CrossRef]

69. Janaki, J.; Mani, A.; Satya, A.T.; Geetha Kumariy, T.; Kalavathi, S.; Bharathi, A. Influence of Ni doping on the electrical and structural properties of $FeSb_2$. *Phys. Status Solidi B* **2012**, *249*, 1756–1760. [CrossRef]

70. Uher, C.; Hu, S.; Yang, J. Cerium filling and lattice thermal conductivity of skutterudites. In Proceedings of the 17th International Conference on Thermoelectrics, Nagoya, Japan, 28 May 1998; pp. 306–309.

71. Kim, H.-S.; Gibbs, Z.M.; Tang, Y.; Wang, H.; Snyder, G.J. Characterization of Lorenz number with Seebeck coefficient measurement. *APL Mater.* **2015**, *3*, 041506. [CrossRef]

72. Aversano, F.; Branz, S.; Bassani, E.; Fanciulli, C.; Ferrario, A.; Boldrini, S.; Baricco, M.; Castellero, A. Effect of rapid solidification on the synthesis and thermoelecrtric properties of Yb-filled $Co_4Sb_{12}$ skutterudite. *J. Alloy Compd.* **2019**, *796*, 33–41. [CrossRef]

 *materials*

*Article*

# Effects of Preparation Procedures and Porosity on Thermoelectric Bulk Samples of Cu₂SnS₃ (CTS)

Ketan Lohani [1], Carlo Fanciulli [2] and Paolo Scardi [1,*]

[1] Department of Civil, Environmental & Mechanical Engineering, University of Trento, Via Mesiano 77, 38123 Trento, Italy; ketan.lohani@unitn.it

[2] Lecco Unit, National Research Council of Italy–Institute of Condensed Matter Chemistry and Technologies for Energy (CNR-ICMATE), Via Previati 1/E, 23900 Lecco, Italy; carlo.fanciulli@cnr.it

* Correspondence: paolo.scardi@unitn.it

**Abstract:** The thermoelectric behavior and stability of Cu₂SnS₃ (CTS) has been investigated in relation to different preparations and sintering conditions, leading to different microstructures and porosities. The studied system is CTS in its cubic polymorph, produced in powder form via a bottom-up approach based on high-energy reactive milling. The as-milled powder was sintered in two batches with different synthesis conditions to produce bulk CTS samples: manual cold pressing followed by traditional sintering (TS), or open die pressing (ODP). Despite the significant differences in densities, ~75% and ~90% of the theoretical density for TS and ODP, respectively, we observed no significant difference in electrical transport. The stable, best performing TS samples reached $zT$ ~0.45, above 700 K, whereas $zT$ reached ~0.34 for the best performing ODP in the same conditions. The higher $zT$ of the TS sintered sample is due to the ultra-low thermal conductivity ($\kappa$ ~0.3–0.2 W/mK), three-fold lower than ODP in the entire measured temperature range. The effect of porosity and production conditions on the transport properties is highlighted, which could pave the way to produce high-performing TE materials.

**Keywords:** copper tin sulfide; Cu₂SnS₃; CTS; thermal stability; chalcogenide; material production; porosity; porous thermoelectric materials

**Citation:** Lohani, K.; Fanciulli, C.; Scardi, P. Effects of Preparation Procedures and Porosity on Thermoelectric Bulk Samples of Cu₂SnS₃ (CTS). *Materials* **2022**, *15*, 712. https://doi.org/10.3390/ma15030712

Academic Editor: Andres Sotelo

Received: 13 December 2021
Accepted: 10 January 2022
Published: 18 January 2022

**Publisher's Note:** MDPI stays neutral with regard to jurisdictional claims in published maps and institutional affiliations.



## 1. Introduction

Thermoelectric (TE) materials are a class of functional materials employed in solid-state energy generators and coolers, without the use of fossil fuels and combustion processes, and with no moving parts, exploiting only the temperature gradients. For this reason, TE materials were first successfully utilized in deep-space probes in the second half of the last century [1]. In recent years, TE materials have drawn attention as efficient, eco-friendly, durable, noise-free, and scalable solutions for the recovery of waste heat, like in exhaust systems (automotive TE generators) [2]. Current research on TE materials seeks greater efficiency, as well as non-toxic and cost-effective solutions [3]. The performance of a TE material can be assessed by means of the figure of merit, $zT = (S^2/\rho\kappa)T$, where $S$, $\rho$, $\kappa$, and $T$ are the Seebeck coefficient, electrical resistivity, thermal conductivity, and absolute temperature, respectively. Thus, a high $zT$ TE material requires a high power factor ($S^2/\rho$) and low $\kappa$, which consists of lattice ($\kappa_l$) and electronic ($\kappa_e$) contributions. That being said, it is not the only important parameter as the stability and durability of TE materials are also vital for producing reliable TE devices.

Cu-Sn-S-based systems are studied as both $p$-type and $n$-type semiconductors [4,5]. CTS is a tetrahedrally bonded direct bandgap semiconductor, showing a bandgap ~1.0 eV, which is higher or lower depending on whether the CTS polymorph is ordered or disordered [6]. The unfilled Cu-$3d$ orbitals introduce holes in the CTS system, making it a $p$-type semiconductor. CTS components are earth-abundant, eco-friendly, non-toxic, non-hazardous, and have low formation energies; therefore, they are explored for various

57

applications, such as TE [7], photovoltaic [8], optoelectronics [9], photoelectrochemical [10], and supercapacitors [11], just to name a few. CTS has attracted the TE community for its remarkable crystallographic structures with different degrees of cation-disorder, which helps in the suppression of thermal conductivity to an ultra-low level, without much hindrance to the electronic transport, also known as the so-called "phonon-glass-electron-crystal (PGEC)" behavior. The well-known ordered or monoclinic CTS polymorph arranges its atoms in the *Cc* space group (SG). In the past, we reported the production of a pure disordered or cubic CTS polymorph (SG: *F-43m*) via high-energy reactive ball-milling [4]. Recently, Wei et al. [12] reported the production of another disordered CTS polymorph, i.e., pure tetragonal (SG: *I-42m*) CTS by the colloidal method.

Interestingly, both the disordered CTS polymorphs (cubic and tetragonal) present $\kappa_l$ <0.5 W/m-K, above 700 K, approaching its theoretical minimum ($\kappa_l$ ~0.3 W/m-K). In a recent report, Li et al. [13] showed the introduction of dislocations as an effective way to suppress the lattice thermal conductivity in Ni-doped CTS ($Cu_2Ni_{0.1}Sn_{0.9}S_3$), reporting $\kappa_l = 0.41$ W/m-K at 723 K. Wei et al. [12] demonstrated that the abundance of twin-boundaries in pristine and In-doped CTS ($Cu_2Sn_{0.95}In_{0.05}S_3$) contributes to a low $\kappa_l = 1.02$ W/m-K at room temperature (RT), which further reduces to $\kappa_l < 0.50$ W/m-K around 700 K. It is important to notice that most materials presenting a high $zT$ in the medium temperature range (500 K < $T$ < 800 K) are either scarce or toxic, e.g., PbTe and SnSe. In this temperature range, sulphides of copper and tin or copper and iron can be viable alternatives. Gu et al. [14] discussed 3D modulation doping of $CuCo_2S_4$ in CTS and reported a $zT$ ~0.82 at 773 K. Zhao et al. [15] reported simultaneous Co and Sb substitution in the CTS system resulting in a $zT$ of ~0.88 at 773 K. Furthermore, Deng et al. [16] reported a $zT$ of ~1 in a similar $Cu_7Sn_3S_{10}$ system through Br-doping. A $zT$ of ~0.4–0.8 above 700 K is regularly reported for CTS by In [17], Zn [7], Mn [18], Ni [19], Fe [20], Co [21], and Cu [22] substitution at the Sn site, aiming to enhance the carrier concentration and reduce the thermal conductivity. Other similar systems, such as $Cu_5Sn_2S_7$ [23], $CuFeS_2$ [24], $Cu_4Sn_7S_{16}$ [25], and $Cu_2SnSe_3$ [26] also show a moderate $zT$ ranging from 0.2–0.5, above 700 K.

Manipulation of porosity in materials is a less-explored subject for TE applications. Many articles have put forward a crucial need to investigate porous TE materials [3,27–29]. Porous materials are lightweight and require lower quantities of materials, therefore, they are suitable for portable and wearable TE devices [27]. Xu et al. [30] showed $zT > 1$, and 2–4 fold lower $\kappa_l$ in hollow nanostructured $Be_2Te_{2.5}Se_{0.5}$ than in the existing literature. Hong et al. [31] showed improved PF in mesoporous ZnO thin films. Tiwari et al. [32] investigated porous CTS for photovoltaics using a solid-state reaction at 200 °C, but they only reported the Seebeck coefficient near RT and temperature-dependent conductivity measurement. Thermal cycling during the TE power generation process could promote the annealing of defects, reduction of grain boundaries, increased density, and transformation of crystal structures, driving these materials towards instability. The authors discussed surface oxidation, the role of secondary phases, and the loss of a nano and microstructure after thermal cycling above a certain temperature, which alters the TE properties of a material [33]. This makes the technological advancement and scientific understanding of these materials challenging [34]. In particular for large-scale production and durable use, a mechanically and thermodynamically stable material is required.

Herein, we stabilized the disordered CTS polymorph using different production conditions and sintering techniques. CTS samples from two batches (three samples each) were produced, and their TE properties and thermal stability were systematically investigated by repeated Seebeck and resistivity measurements (heating and cooling) in the temperature range of 323–723 K. Pellets of samples were produced using manual cold pressing followed by sintering in a traditional tubular furnace (TS batch), or via open die pressing (ODP batch) [35]. Information on the crystallographic structures and phase purity was obtained by Rietveld refinement of X-ray diffraction (XRD) patterns, before and after the repeated TE measurements. Surface topology, porosity, and chemical variations were investigated by morphological images and energy dispersive X-ray (EDX) spectroscopy, using both scanning electron microscopy (SEM) and transmission electron microscopy (TEM). This work provides a general overview and highlights the cautions to produce stable CTS and similar chalcogenides. The effect of porosity on TE properties is highlighted.

## 2. Materials and Methods

CTS samples were produced from the elemental powders using a high-energy reactive ball-mill (Fritsch, Idar-Oberstein, Germany), according to a well-assessed procedure with details discussed elsewhere [4]. Three different samples were prepared using the manual cold press (load 15 tons for 3 min) and sintered at 500 °C for 2 h with a heating rate of 3 °C/min. Sample TS1 was processed in an open environment, whereas samples TS2 and TS3 were produced in a highly controlled environment ($O_2$ and $H_2O$ < 10 ppm). All the samples were sintered in Ar flux. Additionally, sample TS3 was sintered in the presence of sulfur powder (1 g).

The other three samples (ODP1, ODP2, and ODP3) were produced in a controlled environment with the above-mentioned preconditions. Differently, the sintering was performed via ODP using an iron die. To prevent possible reactions between the die and the powders, a boron nitride layer was used to cover the internal surface of the die. Such a layer also prevents sticking phenomena, allowing for an easier extraction of the sintered bulk. Sample ODP1 was sintered at 400 °C for 30 min, whereas samples ODP2 and ODP3 were sintered at 500 °C for 10 and 30 min, respectively. During all the sintering processes, a ~100 MPa axial pressure was applied.

It should be emphasized that ODP was performed on a large quantity of the as-milled powder, namely ~30–40 g. The as-sintered samples from ODP are usually large bulk samples, which are cut into the required sample sizes for various measurements (large as-sintered ODP samples are shown in Figure S1). The densities of the TS (TS1, TS2, and TS3) and ODP (ODP1, ODP2, and ODP3) sintered samples were ~75% and ~90% of the theoretical density (see Table 1), respectively.

The structural characterization of TS samples was performed in Bragg–Brentano geometry using a Rigaku PMG (Rigaku PMG, Tokyo, Japan) powder diffractometer, equipped with a Cu $K\alpha$ ($\lambda$ = 1.5406 Å) source and a scintillation counter detector. XRD data on the ODP samples were collected in Bragg–Brentano geometry using a Bruker D8 (Bruker, Billerica, MA, USA) diffractometer equipped with a Co $K\alpha$ ($\lambda$ = 1.7889 Å) source and area linear (1D) detector. After the phase identification, Rietveld refinement [36] was performed using Whole Powder Pattern Modelling (WPPM) macros [37], as implemented in Topas 7 [38] software (Coelho Software, Brisbane, Australia).

Simultaneous Seebeck coefficient and resistivity measurements were performed in four contact set-up using LSR-3 (Linseis Messgeraete GmbH, Selb, Germany), at a temperature ranging from 323–723 K. Thermal diffusivity (D) was measured using a LFA-500 (Linseis Messgeraete GmbH, Selb, Germany), equipped with a xenon-flash lamp over a temperature range of 323–723 K. Seebeck and resistivity measurements were performed under an He atmosphere (0.1 bar). However, the diffusivity was measured under a vacuum (~$10^{-3}$ bar).

After the TE measurements, morphological images of the selected area electron diffraction (SAED) were collected on powder produced by crushing the sample pellets, and EDX spectroscopy was performed using transmission electron microscopy (TEM; HR-S/TEM ThermoFischer TALOS 200 s, Thermo Fischer Scientific, Waltham, MA, USA). Moreover, morphological images and EDX data were collected on pellet samples to investigate the microstructure and bulk chemistry using a Jeol IT300 scanning electron microscope (SEM) equipped with a tungsten source (Jeol, Ltd., Tokyo, Japan).

**Table 1.** Quantitative phase and average crystallite size before and after TE measurements, measured densities, and relative densities.

| Sample | Weight ($\pm$1%) | | | | | Average Domain Size ($\pm$10 nm) | Measured Density ($\pm$0.1 g/cm$^3$) | Relative Density (%) |
| --- | --- | --- | --- | --- | --- | --- | --- | --- |
| | CTS | WC | SnO$_2$ | SnO | SnS | | | |
| TS1 before TE measurements | 96 | 2 | 2 | - | - | 40 | 3.52 | 74.7 |
| TS1 after TE measurements | 88 | 2 | 8 | - | 2 | 75 | | |
| TS2 before TE measurements | 98 | 2 | - | - | - | 75 | 3.48 | 73.8 |
| TS2 after TE measurements | 89 | 2 | 8 | 1 | - | 75 | | |
| TS3 before TE measurements | 98 | 2 | - | - | - | 70 | 3.68 | 78.1 |
| TS3 after TE measurements | 87 | 2 | 10 | 1 | - | 70 | | |
| ODP1 before TE measurements | 100 | - | - | - | - | 42 | 4.26 | 90.4 |
| ODP1 after TE measurements | 100 | - | - | - | - | 42 | | |
| ODP2 before TE measurements | 100 | - | - | - | - | 54 | 4.36 | 92.5 |
| ODP2 after TE measurements | 99 | 1 | - | - | - | 54 | | |
| ODP3 before TE measurements | 98 | 1 | 1 | - | - | 68 | 4.43 | 94.0 |
| ODP3 after TE measurements | 91 | 1 | 8 | - | - | 68 | | |

## 3. Results and Discussions

XRD measurements were performed on all of the polycrystalline samples (Figure 1). The diffraction patterns showed fingerprint peaks at $2\theta$~28.5°, 47.3°, and 56.0°, and $2\theta$ ~33.2°, 55.5°, and 66.3° using Cu $K\alpha$ and Co $K\alpha$ X-ray sources, respectively. These Bragg peaks represent planes (1 1 1), (2 2 0), and (3 1 1), respectively, confirming the disordered cubic CTS structure, which is derived from the zinc-blende structure (SG: $F$-$43m$). A small amount of WC (SG: $P$-$6m2$) is present in all of the samples due to the use of WC vials for milling. Additionally, a weak peak ($2\theta$ ~27°) of SnO$_2$ (SG: $P_{42}mnm$) was observed in the pattern of sample TS1 (see Figure S2). XRD patterns of all the ODP sintered samples also showed Bragg peaks representing the disordered cubic CTS phase. Nevertheless, small amounts of WC and SnO/SnO$_2$ phases were also observed in some ODP sintered samples.

Different crystallographic phases were identified, and their respective weight fractions were obtained by the Rietveld method [36,39]. Information provided by powder pattern refinements, including the average crystallite size, is reported in Table 1. The lattice parameters for all the samples are $a$ = 5.43 $\pm$ 0.01 Å and $\alpha = \beta = \gamma = 90°$. The average crystallite size for samples TS1, TS2, and TS3 is 40 $\pm$ 10, 75 $\pm$ 10, and 70 $\pm$ 10 nm, respectively. All the samples have a ~1–2% weight of WC. Additionally, sample TS1 has a small quantity (~2% weight) of SnO$_2$, which is possibly formed due to open environment processing. We did not observe the formation of secondary phase oxides in samples TS2 and TS3, produced in a strictly controlled environment. ODP, like spark plasma sintering (SPS), is a fast-sintering technique needing a short sintering time and applying external pressure. The applied pressure during ODP decreases the sintering temperature and time, simultaneously improving the densification kinetics. Thus, the crystalline domains show limited growth, while reaching full densification. The average domain sizes for the ODP1, ODP2, and ODP3 samples are $42 \pm 10$, $54 \pm 10$, and $64 \pm 10$ nm, respectively. In our recent work, we showed that SPS on as-milled CTS powder can constrain the grain growth below 50 nm [40].

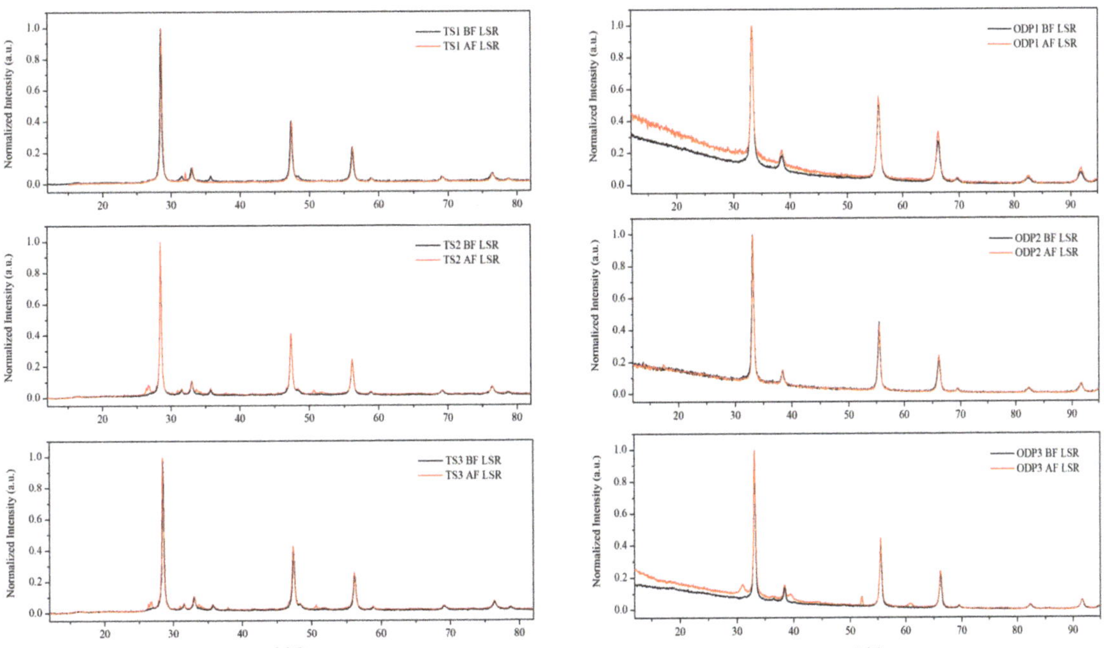

**Figure 1.** XRD data before and after repeated TE measurements on TS and ODP sintered samples. The difference in diffuse background and shifts in Bragg peaks between TS and ODP samples is due to the use of different diffractometers equipped with Cu $K\alpha$ ($\lambda = 1.5406$ Å) and Co $K\alpha$ ($\lambda = 1.7889$ Å) sources, respectively. Rietveld refinement on the XRD data with marked phases is shown in Figure S2.

After the repeated TE measurements (discussed in the next section), the XRD data were collected for the second time. All TS samples showed an increase in the weight fraction of $SnO_2$. Samples TS2 and TS3 showed the formation of a new secondary phase, SnO (SG: $Pnm2_1$), possibly due to an oxygen-deficient production environment. We also observed the presence of SnS (SG: $Pbmn$) in sample TS1, confirming the chemical instability of the sample prepared in an open environment. Samples sintered using ODP were less prone to growth in the secondary phases. We did not observe any grain growth or increase in the weight fraction of secondary phases for ODP1 and ODP2. The applied pressure during sintering seems to better stabilize the samples than traditional sintering. Nevertheless, an increased amount of $SnO_2$ was observed for ODP3, due to its extended sintering time in comparison to ODP2. The densities of ODP3 and ODP2 samples are similar, in spite of the different sintering times of 30 and 10 min, respectively. Therefore, the densification was complete within ~10 min of sintering. Furthermore, extended sintering mainly promotes phase degradation, leading to the formation of $SnO_2$.

As discussed in the Introduction, the loss of nanostructure has adverse effects on the TE properties of the material, which could drive it towards instability. The average crystalline size for the TS1 sample increased from $40 \pm 10$ to $75 \pm 10$ nm after repeated TE measurements, whereas samples TS2 and TS3 did not lose their nanostructure, maintaining their average crystallite size of $75 \pm 10$ and $70 \pm 10$ nm, respectively. Similarly, the ODP sintered samples maintained their average domain size of $42 \pm 10$, $54 \pm 10$, and $68 \pm 10$ nm for ODP1, ODP2, and ODP3 samples, respectively, after repeated TE measurements.

SEM micrographs at different magnifications collected on TS and ODP sintered sample surfaces are shown in Figure 2. The difference between the two samples is noticeable. Morphological images of the bulk TS samples (Figure 2a,b) reveal an uneven topology and the presence of microscopic pores. As expected, the ODP sintered samples (Figure 2c,d) had a lower porosity compared to TS, with intact grains. It is also evident through the density measurements that the ODP sintered samples were ~20% denser than TS. Here, it is worth mentioning that similar samples sintered via SPS showed almost no porosity and the density was slightly (~2–4%) higher than the ODP sintered samples (morphological images of SPS sintered samples are shown in Figure S3) [40].

The morphological image and SAED using TEM on TS1, shown in Figure S4 (right bottom inset), reveals many small (d~10 nm) grains surrounding the bigger (CTS) grains. EDX was performed, focusing on the surrounding small dark grains. Strong peaks of oxygen and tin can be observed in the EDX spectra of TS1, and the atomic fraction was found to be oxygen and tin-rich (see Table 2), suggesting that these are $SnO_2$ grains, as we observed from the XRD. The evolution of $SnO_2$ grains makes the sample unstable during TE measurements by unbalancing the stoichiometry. For samples TS2 and TS3, no $SnO_2/SnO$ grains appeared in the micrographs, although a weak peak for oxygen could be observed by EDX. The chemical composition for the TS2 sample was copper-rich, whereas, for the sulfurized sample (TS3), the stoichiometry was close to the theoretical (see Table 2).

**Table 2.** Atomic fraction for samples TS1, TS2, and TS3 estimated from the TEM-EDX analysis (corresponding EDX spectra are shown in Figure S4).

| Sample Name | Atomic Fraction ($\pm 1$%) | | | |
|---|---|---|---|---|
| | Cu | Sn | S | O |
| TS1 | 15.11 | 26.40 | 14.92 | 43.55 |
| TS2 | 40.09 | 16.64 | 41.55 | 1.7 |
| TS3 | 30.31 | 15.43 | 45.96 | 3.2 |

SEM-EDX analysis was performed on the ODP sintered samples, including chemical maps on large ODP samples to investigate the overall chemical homogeneity. As discussed in the Experimental Methods, ODP was performed on a large quantity (30–40 g) of ball-milled powder, producing large samples (see Figure S1). The SEM chemical maps (shown in Figure 3) revealed non-homogeneous chemistry on the surfaces of the ODP sintered samples. As ODP is a fast-sintering technique and was performed on a large quantity of materials, variations in chemical composition are likely. However, the overall chemistry of the samples was close to theoretical (shown in Table 3). On the contrary, the SPS sintered samples produced in our past work showed a homogeneous chemical distribution and comparatively lower crystalline domain size [40].

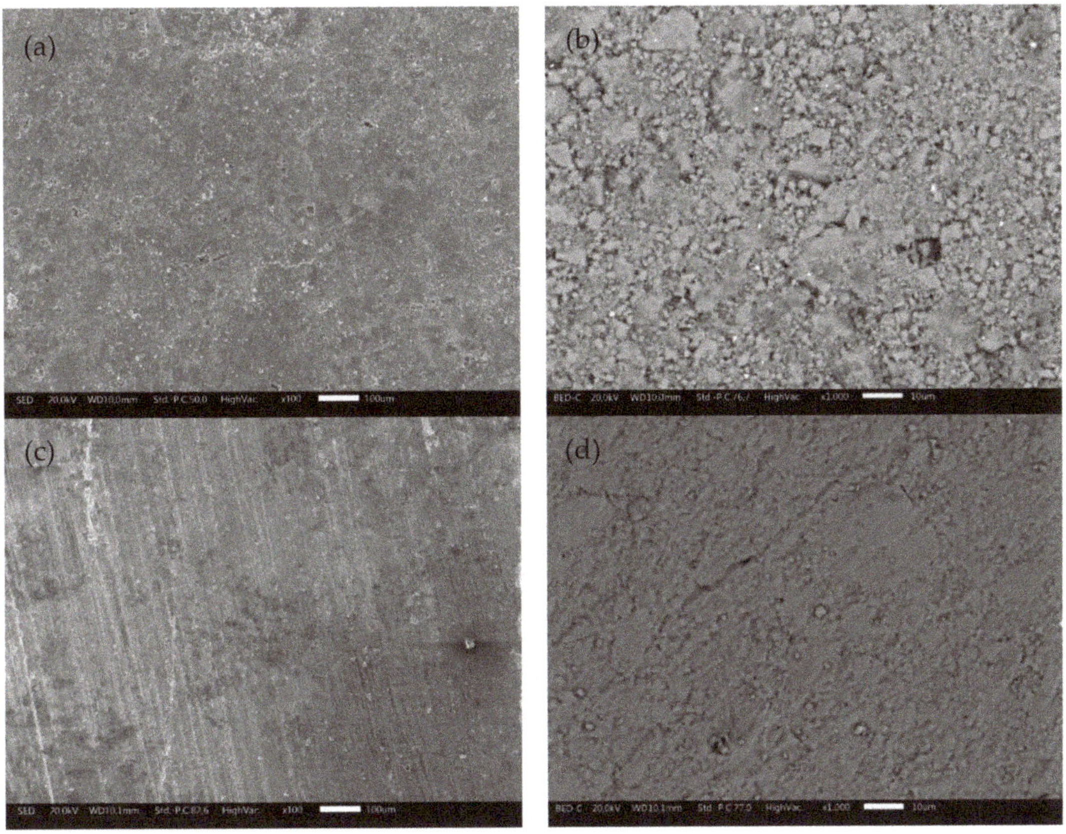

**Figure 2.** SEM micrographs collected on the surfaces of bulk TS (**a**,**b**) and ODP (**c**,**d**) sintered samples showing highly intact grains with a lower porosity in comparison with the TS sintered samples at different magnifications.

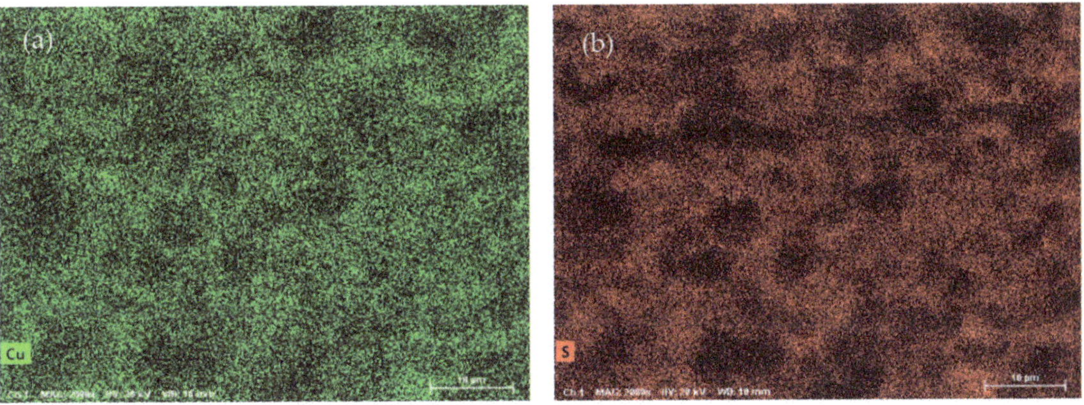

**Figure 3.** *Cont.*

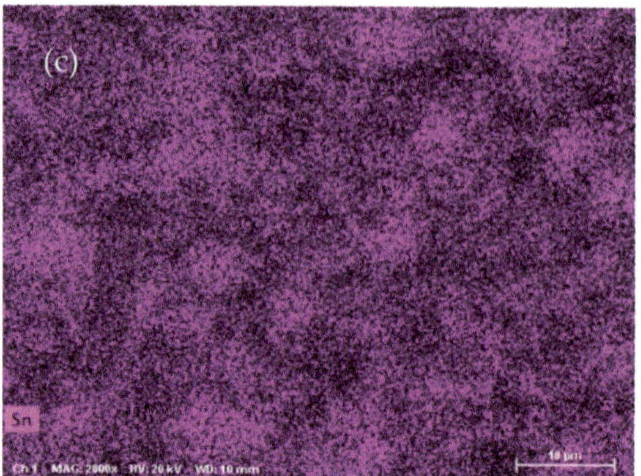
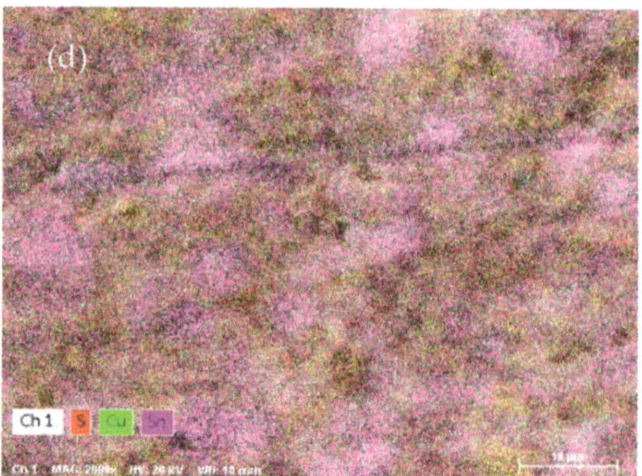

**Figure 3.** SEM micrographs collected on ODP samples and chemical maps for individual elements Cu (**a**), S (**b**), and Sn (**c**), and simultaneously for all elements (**d**) (corresponding EDX spectra are shown in Figure S5).

**Table 3.** Atomic fraction of samples ODP1, ODP2, and ODP3 samples, estimated using SEM-EDX analysis (corresponding EDX spectra are shown in Figure S6).

| Sample Name | Atomic Fraction (±1%) | | | |
|---|---|---|---|---|
| | Cu | Sn | S | O |
| ODP1 | 16.90 | 32.03 | 46.92 | 4.05 |
| ODP2 | 15.46 | 30.13 | 45.11 | 9.27 |
| ODP3 | 16.82 | 31.68 | 49.21 | 2.29 |

Repeated Seebeck and resistivity measurements are shown in Figure 4. All the samples, and TS and ODP batches, showed positive values for the absolute Seebeck coefficient, indicating holes as majority charge carriers, thus confirming a $p$-type semiconducting nature. The trend of resistivity for sample TS1 is increasing with the temperature, whereas it is decreasing for all other samples, typical of degenerate and non-degenerate semiconductors, respectively.

The first measurement on the TS1 sample shows the lowest values of $S$ (40–100 μV/K) and $\rho$ (0.01–0.03 mΩ-m). However, these measurements are not reproducible during the second and third measurement cycles. The low Seebeck and resistivity with a non-degenerate semiconducting trend suggest the presence of a higher carrier concentration in sample TS1. This could be due to the alteration in chemical composition, as we observed a large amount of $SnO_2$ using TEM-EDX and XRD. The XRD on the same sample after repeated TE measurement cycles revealed the segregation of SnS (see Table 1), making the sample copper-rich or possibly leading to the formation of CuS. In a recent work on non-stoichiometric (copper-rich) CTS [18], lower Seebeck (~50 μV/K at 300 K) and high electrical conductivity were reported. Other than the formation of SnS and $SnO_2$, the non-repeatability of $S$ and $\rho$ during the thermal cycles also confirms the instability of sample TS1. Noticeably, after the three measurements cycles, significant grain growth was observed for TS1, with the average domain size increasing from 40 ± 10 to 75 ± 10 nm.

For samples TS2 ($S$ ~250–400 μV/K, $\rho$ ~1.5–1.1 mΩ-m) and TS3 ($S$ ~325–425 μV/K, $\rho$ ~3.5–1.5 mΩ-m), the TE measurements were repeatable during the measurement cycles, and the values of $S$ and $\rho$ were in agreement with the literature [7,15]. Although we observed an increased amount of $SnO_2$ and the formation of a new SnO phase, segregation of SnS was not observed. Similar values of $S$ and $\rho$ during several measurement cycles confirmed that the CTS samples prepared in a strictly controlled environment were stable.

It is known that sulfurization leads to the formation of stoichiometric CTS [41], which could be the case for sample TS3, leading to a higher Seebeck and resistivity.

The repeated TE measurements on ODP sintered samples were reproducible, especially after the first measurement cycle, and for the second and third measurements cycles, the $S$ and $\rho$ measurements were coherent. The ODP1 sample sintered at 400 °C for 30 min showed the lowest values of $S$ (180–310 µV/K) and $\rho$ (0.55–0.50 mΩ-m). ODP2 and ODP3 samples sintered at 500 °C for 10 min and 30 min showed increasing values of $S$ and $\rho$. The Seebeck coefficients and resistivities for the ODP2 and ODP3 samples were $S$ ~400–550 µV/K and $\rho$ ~2.2–1.0 mΩ-mm, and $S$ ~550–650 µV/K and $\rho$ ~4.0–1.7 mΩ-m, respectively. The increase in $S$ and $\rho$ have a clear relation with the increased average domain size of the samples. These results are in agreement with our recent work showing the effects of grain growth on SPS sintered CTS samples, owing to the conduction based on the metallic nature of surfaces due to dangling bonds [40]. Moreover, Ming et al. [42] showed a similar effect in $Cu_2SnSe_3$.

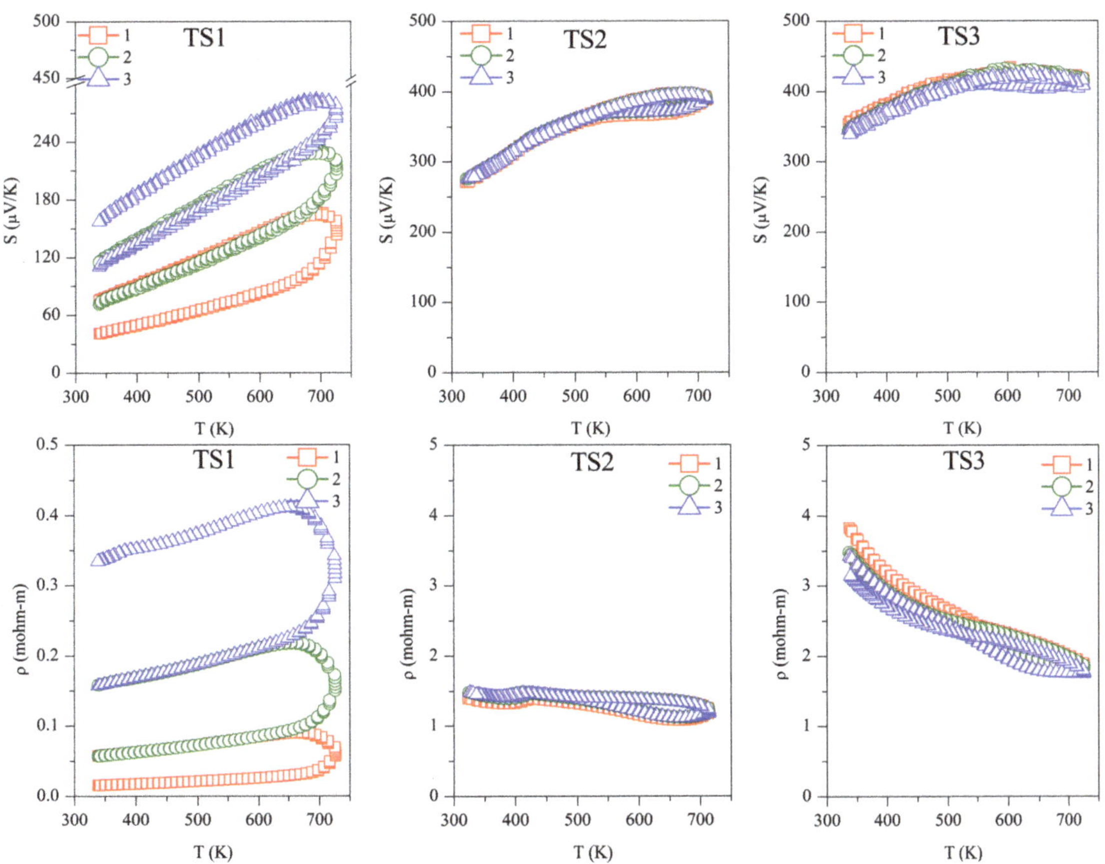

**Figure 4.** *Cont.*

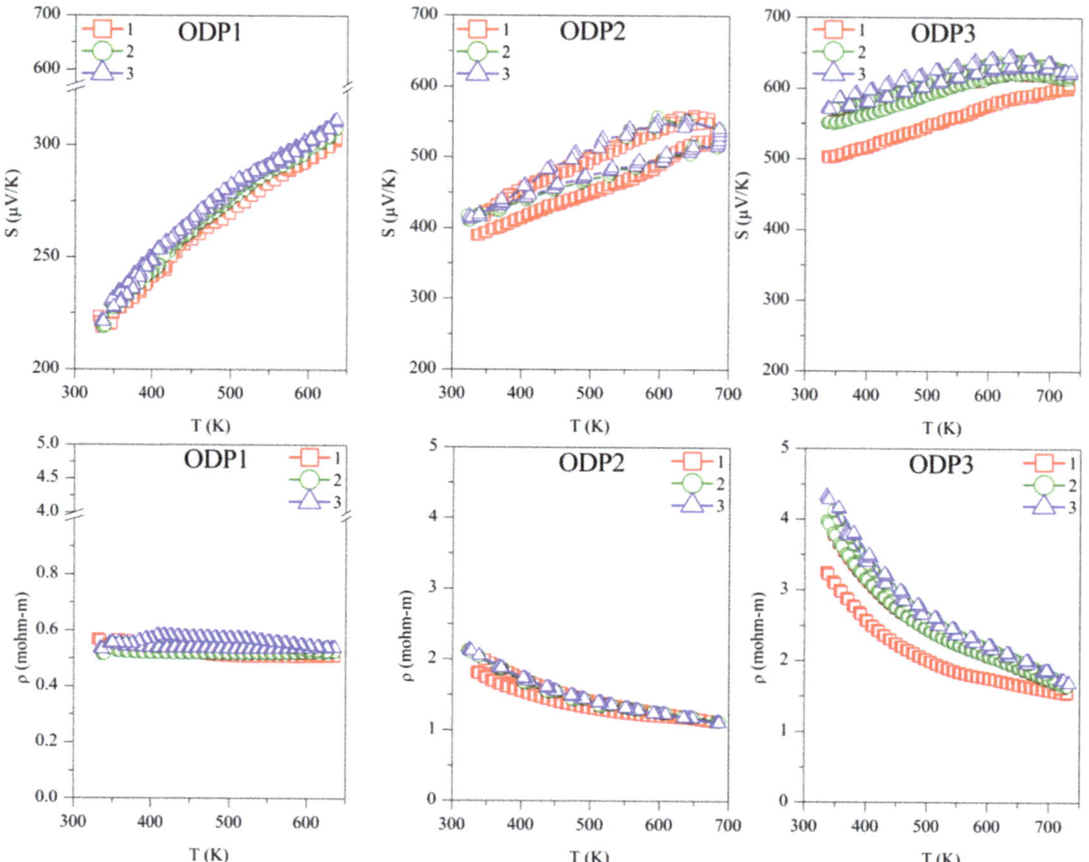

**Figure 4.** Repeated absolute Seebeck coefficient ($S$) and resistivity ($\rho$) measurements in temperature for the TS and ODP sintered samples. In the figure, markers 1 (red squares), 2 (green circles), and 3 (blue triangles) represent first, second, and third measurements cycles, respectively.

However, it is important to notice here that two different sintering were used for the sample preparation, resulting in significantly different densities, even if we did not observe a significant difference in the resistivity of the same samples. Thus, we put forward that electrical transport is less likely affected by porosity in these systems. However, we noticed that the TS sintered samples had a lower $S$ in comparison with the ODP samples. This could be related to the overall chemical composition of the TS samples, where we noticed a large chemical fluctuation in TEM-EDX. Some chemical fluctuations were also present in the ODP samples, but the overall chemistry of ODP samples was close to the theoretical, as observed by means of the chemical maps.

Thermal diffusivity ($D$) was measured using a xenon flash instrument, and the thermal conductivity ($\kappa$) was calculated as, $\kappa = D \times d \times C_p$, where, $d$ and $C_p$ are the density of the sample and specific heat capacity, respectively. The density was measured using the Archimedes method (see Table 1), and $C_p$ obtained on TS samples is discussed elsewhere [4], whereas $C_p$ measurements on ODP sintered samples are shown in Figure S7.

All the samples showed a decreasing trend of thermal conductivity with temperature, implying that the phonon–phonon (Umkalpp process) is dominating the phonon transport (see Figure 5). Here, it is worth mentioning that all the samples of this work have a disordered cubic structure with partial occupancies of cations (Cu and Sn), which results in a comparatively lower thermal conductivity than the ordered or monoclinic CTS polymorph [6]. The thermal conductivity of sample TS1 ranges from 0.70 W/mK to 0.45 W/mK, whereas it is ultra-low for samples TS2 and TS3, with values of 0.40–0.30 W/mK and 0.30–0.20 W/mK, respectively, in the entire measured temperature range of 323–723 K. Samples TS2 and TS3 have a similar $\kappa$ as we recently reported for CTS [4]. However, the unstable sample TS1 has the highest $\kappa$ in the TS batch. The higher thermal conductivity of sample TS1 is due to its higher electrical conductivity, as observed by the degenerate semiconductor trend of resistivity, probably resulting from the formation of Cu-rich CTS or CuS. The thermal conductivity of ODP sintered samples is in the range of other CTS and Cu-Sn-S-based systems [7,18]. A clear correlation between densities and $\kappa$ can be noticed in the ODP samples as well. The densities of ODP samples are similar to the literature for CTS polymorphs [7,12,18]. The difference in $\kappa$ between TS2 and TS3 is also likely to be due to their lower and higher densities, respectively. It is noteworthy that the $\kappa$ of TS samples is ~3-fold lower in the entire temperature range than for the ODP sintered samples. However, we observed a more or less same behavior for the electrical transport, e.g., samples TS2 and ODP2 have $\rho$~1.5–1.0 m$\Omega$-m and ~2.0–1.2 m$\Omega$-m, respectively, but TS2 has a ~3-fold lower $\kappa$ (0.3–0.2 W/mK) than ODP2 (0.8–0.6 W/mK).

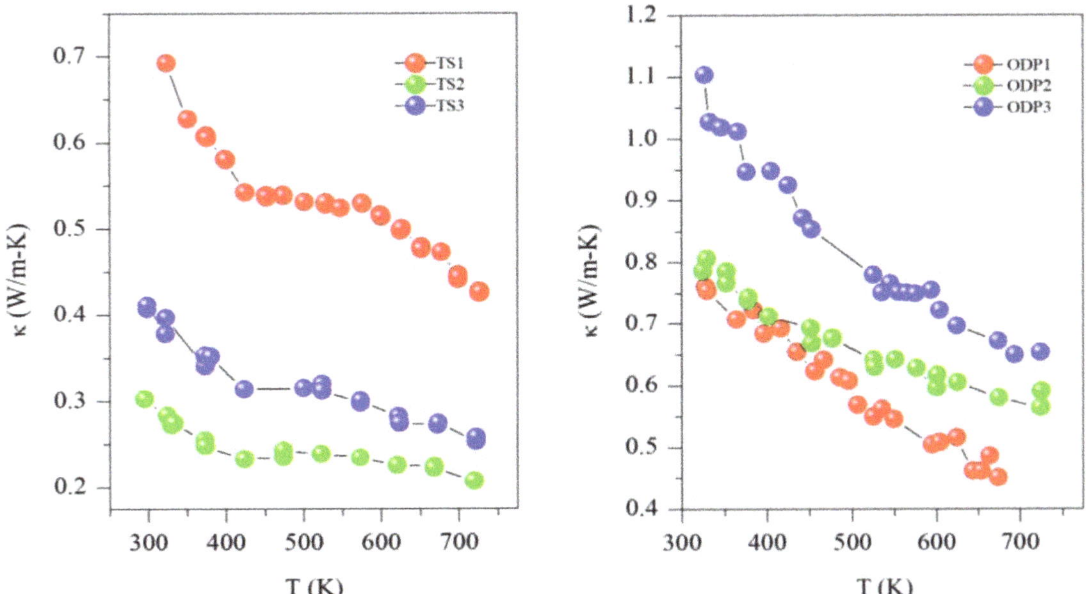


**Figure 5.** Thermal conductivity ($\kappa$) for TS and ODP sintered samples.

The power factor was calculated as $PF = S^2/\rho$, and is shown in Figure 6. The first measurement on Sample TS1 has the highest $PF$ ~3.5 µW/K$^2$ cm, above 700 K, which is continuously decreasing with successive measurements. Samples TS2 and TS3 showed PF ~1.25 µW/K$^2$ cm and ~1 µW/K$^2$ cm, respectively, above 700 K, but they are stable over repeated measurements. Due to the reproducibility of $S$ and $\rho$, ODP sintered samples showed reproducible PF. Among all the ODP sintered samples, ODP2 has the highest $PF$ ~2.75 µW/K$^2$ cm, above 700 K, because of moderate $S$ and $\rho$. However, lower $S$ and very high $\rho$ for samples ODP1 and ODP3, respectively, have an adverse effect on each other, resulting in a comparatively lower $PF$ ~1.75 µW/K$^2$ cm and ~2.25 µW/K$^2$ cm, respectively.

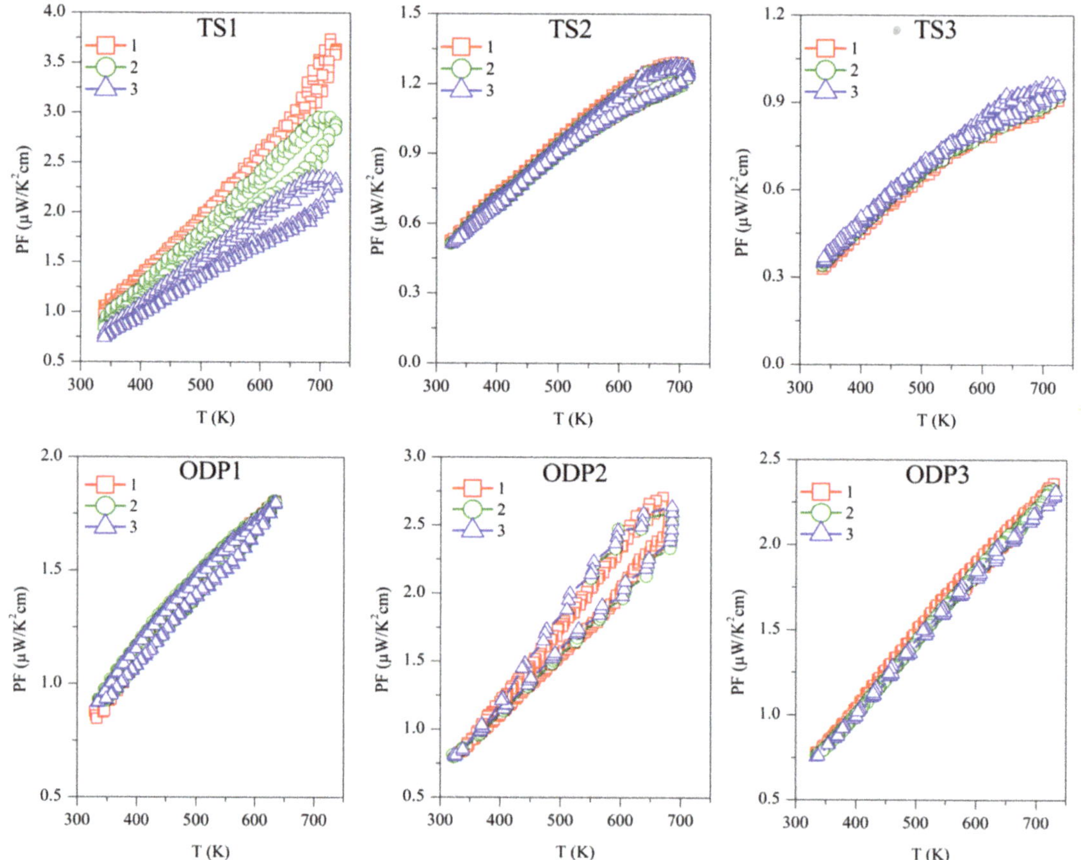

**Figure 6.** Repeated power factor (PF) for TS and ODP sintered samples. In the figure, markers 1 (red squares), 2 (green circles), and 3 (blue triangles) represent the power factors during the first, second, and third measurements cycles, respectively.

Concerning the figure of merit, samples TS1, TS2, and TS3 show $zT$ ~0.50, 0.45, and 0.27, respectively, above 700 K (Figure 7). Although TS1 has the highest $zT$, it is not stable and $zT$ decreases with subsequent measurements. Among the stable TS samples, TS2 has the highest $zT$ of ~0.45. The competitively high $zT$ of TS2 is supported by its lower $\kappa$ and $\rho$. ODP sintered samples present $zT$ ranging from 0.25–0.34, around 700 K. The ODP2 sample has the highest $zT$ of ~0.34, around 700 K, thanks to the low thermal conductivity throughout the measured temperature span. It is evident from the discussions above that the comparatively higher $zT$ of the TS samples is supported by their lower density (high porosity), which blocked the thermal transport effectively. Impotently, the many folds reduction of $\kappa$ due to the pores does not show a significant increase in the resistivity of the TS samples. Furthermore, the porous CTS samples also showed reproducible results during many measurements cycles. Total elimination of secondary phase oxides is not easy due to the low partial pressure of their formations, but a small amount of tin oxides does not seem to affect the TE properties during repeated TE measurements.

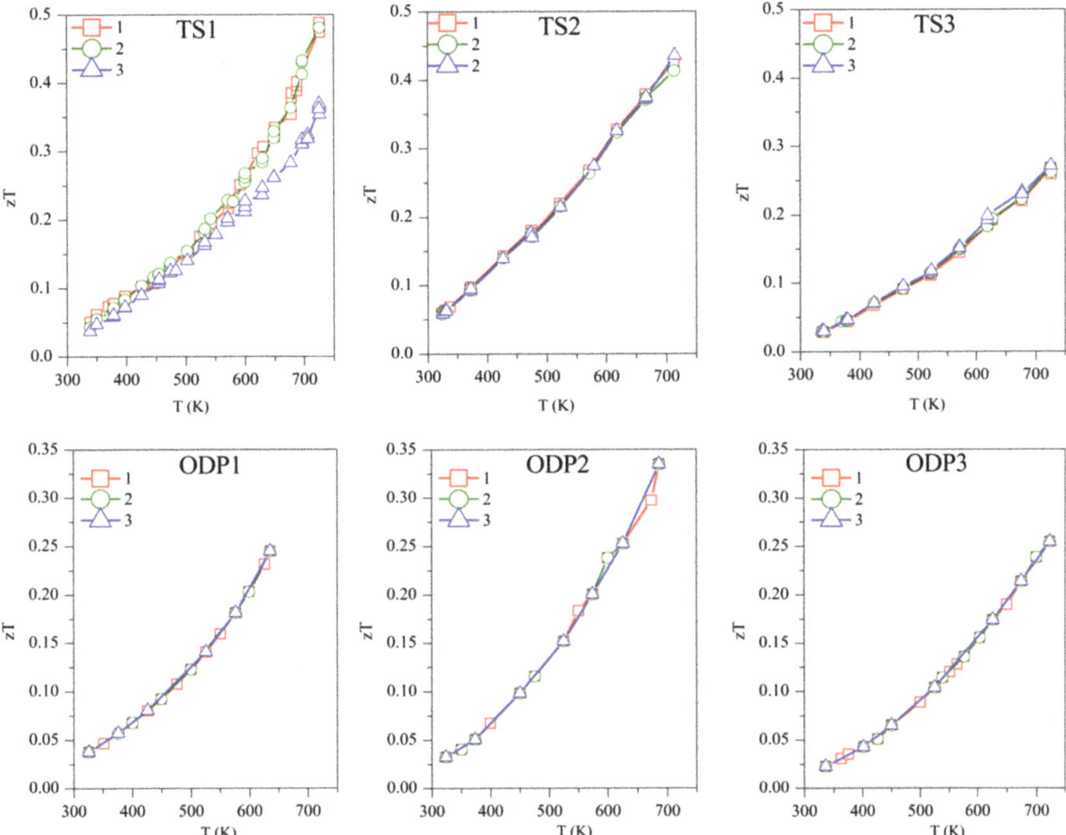

**Figure 7.** The repeated figure of merit ($zT$) for TS and ODP samples. In the figure, markers 1 (red star), 2 (green circles), and 3 (blue triangles) represent the first, second, and third measurement cycles, respectively.

## 4. Conclusions

The present work shows that porosity leads to a significant suppression of thermal conductivity, and CTS pellets can be stabilized with different fractions of porosity. The low density or TS samples ($\kappa$ ~0.3–0.2 W/mK) showed ~3 times lower thermal conductivity than the high density or ODP sintered samples ($\kappa$ ~0.8–0.6 W/mK). However, we did not observe any significant difference in electrical transport properties between the TS and ODP samples. The best performing stable TS samples present $zT$~0.45, whereas the best performing ODP sample showed $zT$ of ~0.34, around 700 K, a result clearly due to the ultra-low thermal conductivity of the traditionally sintered, porous samples.

Due to the low partial pressure of tin oxides formation, it is difficult to produce pure CTS and similar chalcogenides. However, the continuous evolution of secondary phase oxides can be eliminated by using a highly controlled environment, and a small fraction of secondary phase oxides seem to have little effect on the TE properties and stability of the samples. The results presented in this work give a general overview of the effects of different experimental conditions and porosity on the stability and TE performance of CTS samples. Similar considerations should hold for other Cu-Sn/Fe-S/Se-based systems, chalcogenides, colusites, and chalcopyrite, etc., used for various applications, ranging from photovoltaics to thermoelectricity and LED production.

**Supplementary Materials:** The following supporting information can be downloaded at https://www.mdpi.com/article/10.3390/ma15030712/s1. Figure S1: Images of as-sintered ODP samples. Figure S2: Rietveld refinement data before and after many Seebeck and resistivity measurements cycles for all the samples. Figure S3: Morphological images of SPS sintered samples. Figure S4: TEM-EDX, morphological images, and SAED on TE samples. Figure S5: SEM-EDX data for chemical maps collected on ODP samples. Figure S6: Various SEM-EDX data collected on ODP samples. Figure S7: Specific heat capacity ($C_P$) measurements on ODP samples.

**Author Contributions:** Conceptualization and original draft writing: K.L.; sample production: K.L. and C.F.; data collection: K.L. and C.F.; discussions, editing, and proofreading: K.L., C.F. and P.S.; supervision and funding, P.S. All authors have read and agreed to the published version of the manuscript.

**Funding:** This research was funded by the Autonomous Province of Trento, within the framework of the programmatic Energy Action 2015–2017.

**Data Availability Statement:** The data presented in this study are available on request from the corresponding author.

**Acknowledgments:** The authors are thankful to Himanshu Nautiyal for useful discussions and suggestions. The authors are also grateful to Mirco D'Incau and Narges Ataollahi for support in the experimental work and for useful discussions on the methods.

**Conflicts of Interest:** The authors declare no conflict of interest.

# References


1. Snyder, G.J.; Toberer, E.S. Complex thermoelectric materials. *Nat. Mater.* **2008**, *7*, 105–114. [CrossRef] [PubMed]
2. Vining, C.B. An inconvenient truth about thermoelectrics. *Nat. Mater.* **2009**, *8*, 83–85. [CrossRef] [PubMed]
3. Mao, J.; Liu, Z.; Zhou, J.; Zhu, H.; Zhang, Q.; Chen, G.; Ren, Z. Advances in thermoelectrics. *Adv. Phys.* **2018**, *67*, 69–147. [CrossRef]
4. Lohani, K.; Isotta, E.; Ataollahi, N.; Fanciulli, C.; Chiappini, A.; Scardi, P. Ultra-low thermal conductivity and improved thermoelectric performance in disordered nanostructured copper tin sulphide (Cu2SnS3, CTS). *J. Alloys Compd.* **2020**, *830*, 154604. [CrossRef]
5. He, T.; Lin, N.; Du, Z.; Chao, Y.; Cui, J. The role of excess Sn in Cu 4 Sn 7 S 16 for modification of the band structure and a reduction in lattice thermal conductivity. *J. Mater. Chem. C* **2017**, *5*, 4206–4213. [CrossRef]
6. Lohani, K.; Nautiyal, H.; Ataollahi, N.; Fanciulli, C.; Sergueev, I.; Etter, M.; Scardi, P. Experimental and Ab Initio Study of Cu 2 SnS 3 (CTS) Polymorphs for Thermoelectric Applications. *J. Phys. Chem. C* **2021**, *125*, 178–188. [CrossRef]
7. Shen, Y.; Li, C.; Huang, R.; Tian, R.; Ye, Y.; Pan, L.; Koumoto, K.; Zhang, R.; Wan, C.; Wang, Y. Eco-friendly p-type Cu2SnS3 thermoelectric material: Crystal structure and transport properties. *Sci. Rep.* **2016**, *6*, 32501. [CrossRef]
8. Lokhande, A.C.; Shelke, A.; Babar, P.T.; Kim, J.; Lee, D.J.; Kim, I.C.; Lokhande, C.D.; Kim, J.H. Novel antibacterial application of photovoltaic Cu2SnS3 (CTS) nanoparticles. *RSC Adv.* **2017**, *7*, 33737–33744. [CrossRef]
9. Oliva, F.; Arqués, L.; Acebo, L.; Guc, M.; Sánchez, Y.; Alcobé, X.; Pérez-Rodríguez, A.; Saucedo, E.; Izquierdo-Roca, V. Characterization of Cu 2 SnS 3 polymorphism and its impact on optoelectronic properties. *J. Mater. Chem. A* **2017**, *5*, 23863–23871. [CrossRef]
10. Jathar, S.B.; Rondiya, S.R.; Jadhav, Y.A.; Nilegave, D.S.; Cross, R.W.; Barma, S.V.; Nasane, M.P.; Gaware, S.A.; Bade, B.R.; Jadkar, S.R.; et al. Ternary Cu2SnS3: Synthesis, Structure, Photoelectrochemical Activity, and Heterojunction Band Offset and Alignment. *Chem. Mater.* **2021**, *33*, 1983–1993. [CrossRef]
11. Wang, C.; Tian, H.; Jiang, J.; Zhou, T.; Zeng, Q.; He, X.; Huang, P.; Yao, Y. Facile Synthesis of Different Morphologies of Cu2SnS3 for High-Performance Supercapacitors. *ACS Appl. Mater. Interfaces* **2017**, *9*, 26038–26044. [CrossRef] [PubMed]
12. Wei, Y.; Zhou, Z.; Jiang, P.; Zheng, S.; Xiong, Q.; Zhang, B.; Wang, G.; Lu, X.; Han, G.; Zhou, X. Phase Composition Manipulation and Twin Boundary Engineering Lead to Enhanced Thermoelectric Performance of Cu 2 SnS 3. *ACS Appl. Energy Mater.* **2021**, *4*, 9240–9247. [CrossRef]
13. Li, C.; Song, H.; Cheng, Y.; Qi, R.; Huang, R.; Cui, C.; Wang, Y.; Zhang, Y.; Miao, L. Highly Suppressed Thermal Conductivity in Diamond-like Cu 2 SnS 3 by Dense Dislocation. *ACS Appl. Energy Mater.* **2021**, *4*, 8728–8733. [CrossRef]
14. Gu, Y.; Ai, W.; Zhao, Y.; Hu, X.; Pan, L.; Zong, P.; Lu, C.; Xu, Z.; Wang, Y. Remarkable thermoelectric property enhancement in Cu2SnS3-CuCo2S4 nanocomposites via 3D modulation doping. *J. Mater. Chem. A* **2021**, *9*, 1928–16935. [CrossRef]
15. Zhao, Y.; Gu, Y.; Zhang, P.; Hu, X.; Wang, Y.; Zong, P.; Pan, L.; Lyu, Y.; Koumoto, K. Enhanced thermoelectric performance in polymorphic heavily Co-doped Cu 2 SnS 3 through carrier compensation by Sb substitution. *Sci. Technol. Adv. Mater.* **2021**, *22*, 363–372. [CrossRef]
16. Deng, T.; Qiu, P.; Xing, T.; Zhou, Z.; Wei, T.R.; Ren, D.; Xiao, J.; Shi, X.; Chen, L. A low-cost and eco-friendly Br-doped Cu7Sn3S10thermoelectric compound withzTaround unity. *J. Mater. Chem. A* **2021**, *9*, 7946–7954. [CrossRef]

17. Tan, Q.; Sun, W.; Li, Z.; Li, J.F. Enhanced thermoelectric properties of earth-abundant Cu2SnS3 via In doping effect. *J. Alloys Compd.* **2016**, *672*, 558–563. [CrossRef]
18. Zhang, Z.; Zhao, H.; Wang, Y.; Hu, X.; Lyu, Y.; Cheng, C.; Pan, L.; Lu, C. Role of crystal transformation on the enhanced thermoelectric performance in Mn-doped Cu2SnS3. *J. Alloys Compd.* **2019**, *780*, 618–625. [CrossRef]
19. Xu, X.; Zhao, H.; Hu, X.; Pan, L.; Chen, C.; Li, D.; Wang, Y. Synergistic role of Ni-doping in electrical and phonon transport properties of Cu2Sn1-xNixS3. *J. Alloys Compd.* **2017**, *728*, 701–708. [CrossRef]
20. Zhao, L.; Chen, C.; Pan, L.; Hu, X.; Lu, C.; Wang, Y. Magnetic iron doping in Cu 2 SnS 3 ceramics for enhanced thermoelectric transport properties. *J. Appl. Phys.* **2019**, *125*, 095107. [CrossRef]
21. Zhao, H.; Xu, X.; Li, C.; Tian, R.; Zhang, R.; Huang, R.; Lyu, Y.; Li, D.; Hu, X.; Pan, L.; et al. Cobalt-doping in Cu2SnS3: Enhanced thermoelectric performance by synergy of phase transition and band structure modification. *J. Mater. Chem. A* **2017**, *5*, 23267–23275. [CrossRef]
22. Deng, T.; Qiu, P.; Song, Q.; Chen, H.; Wei, T.-R.; Xi, L.; Shi, X.; Chen, L. Thermoelectric properties of non-stoichiometric Cu2+xSn1−xS3 compounds. *J. Appl. Phys.* **2019**, *126*, 085111. [CrossRef]
23. Pavan Kumar, V.; Lemoine, P.; Carnevali, V.; Guélou, G.; Lebedev, O.I.; Boullay, P.; Raveau, B.; Al Rahal Al Orabi, R.; Fornari, M.; Prestipino, C.; et al. Ordered sphalerite derivative Cu 5 Sn 2 S 7: A degenerate semiconductor with high carrier mobility in the Cu–Sn–S diagram. *J. Mater. Chem. A* **2021**, *9*, 10812–10826. [CrossRef]
24. Baláž, P.; Dutková, E.; Levinský, P.; Daneu, N.; Kubíčková, L.; Knížek, K.; Baláž, M.; Navrátil, J.; Kašparová, J.; Ksenofontov, V.; et al. Enhanced thermoelectric performance of chalcopyrite nanocomposite via co-milling of synthetic and natural minerals. *Mater. Lett.* **2020**, *275*, 128107. [CrossRef]
25. Deng, T.; Wei, T.R.; Song, Q.; Xu, Q.; Ren, D.; Qiu, P.; Shi, X.; Chen, L. Thermoelectric properties of n-type Cu4Sn7S16-based compounds. *RSC Adv.* **2019**, *9*, 7826–7832. [CrossRef]
26. Siyar, M.; Cho, J.Y.; Youn, Y.; Han, S.; Kim, M.; Bae, S.H.; Park, C. Effect of annealing temperature on the phase transition, band gap and thermoelectric properties of Cu2SnSe3. *J. Mater. Chem. C* **2018**, *6*, 1780–1788. [CrossRef]
27. Giulia, P. Thermoelectric materials: The power of pores. *Nat. Rev. Mater.* **2017**, *2*, 17006. [CrossRef]
28. Goldsmid, H. Porous Thermoelectric Materials. *Materials* **2009**, *2*, 903–910. [CrossRef]
29. Xu, B.; Feng, T.; Li, Z.; Pantelides, S.T.; Wu, Y. Constructing Highly Porous Thermoelectric Monoliths with High-Performance and Improved Portability from Solution-Synthesized Shape-Controlled Nanocrystals. *Nano Lett.* **2018**, *18*, 4034–4039. [CrossRef]
30. Xu, B.; Feng, T.; Agne, M.T.; Zhou, L.; Ruan, X.; Snyder, G.J.; Wu, Y. Highly Porous Thermoelectric Nanocomposites with Low Thermal Conductivity and High Figure of Merit from Large-Scale Solution-Synthesized Bi 2 Te 2.5 Se 0.5 Hollow Nanostructures. *Angew. Chem. Int. Ed.* **2017**, *56*, 3546–3551. [CrossRef]
31. Hong, M.-H.; Choi, H.; Kim, Y.; Shim, D., II; Cho, H.H.; Park, H.-H. Thermoelectric behaviors of ZnO mesoporous thin films affected by strain induced from the different dopants radii (Al, Ga, and In). *Appl. Phys. Lett.* **2021**, *119*, 193902. [CrossRef]
32. Tiwari, D.; Chaudhuri, T.K.; Shripathi, T.; Deshpande, U. Synthesis of earth-abundant Cu2SnS3 powder using solid state reaction. *J. Phys. Chem. Solids* **2014**, *75*, 410–415. [CrossRef]
33. Aversano, F.; Palumbo, M.; Ferrario, A.; Boldrini, S.; Fanciulli, C.; Baricco, M.; Castellero, A. Intermetallics Role of secondary phases and thermal cycling on thermoelectric properties of TiNiSn half-Heusler alloy prepared by different processing routes. *Intermetallics* **2020**, *127*, 106988. [CrossRef]
34. Baranowski, L.L.; Zawadzki, P.; Lany, S.; Toberer, E.S.; Zakutayev, A. A review of defects and disorder in multinary tetrahedrally bonded semiconductors. *Semicond. Sci. Technol.* **2016**, *31*, 123004. [CrossRef]
35. Fanciulli, C.; Coduri, M.; Boldrini, S.; Abedi, H.; Tomasi, C.; Famengo, A.; Ferrario, A.; Fabrizio, M.; Passaretti, F. Structural texture induced in SnSe thermoelectric compound via open die pressing. *J. Nanosci. Nanotechnol.* **2017**, *17*, 1571–1578. [CrossRef]
36. McCusker, L.B.; Von Dreele, R.B.; Cox, D.E.; Louër, D.; Scardi, P. Rietveld refinement guidelines. *J. Appl. Crystallogr.* **1999**, *32*, 36–50. [CrossRef]
37. Scardi, P.; Azanza Ricardo, C.L.; Perez-Demydenko, C.; Coelho, A.A. Whole powder pattern modelling macros for TOPAS. *J. Appl. Crystallogr.* **2018**, *51*, 1752–1765. [CrossRef]
38. Coelho, A.A. TOPAS and TOPAS-Academic: An optimization program integrating computer algebra and crystallographic objects written in C++. *J. Appl. Crystallogr.* **2018**, *51*, 210–218. [CrossRef]
39. Scardi, P. Diffraction Line Profiles in the Rietveld Method. *Cryst. Growth Des.* **2020**, *20*, 6903–6916. [CrossRef]
40. Lohani, K.; Nautiyal, H.; Ataollahi, N.; Maji, K.; Guilmeau, E.; Scardi, P. Effects of Grain Size on the Thermoelectric Properties of Cu2SnS3: An Experimental and First-Principles Study. *ACS Appl. Energy Mater.* **2021**, *4*, 12604–12612. [CrossRef]
41. Dong, Y.; He, J.; Li, X.; Zhou, W.; Chen, Y.; Sun, L.; Yang, P.; Chu, J. Synthesis and optimized sulfurization time of Cu2SnS3 thin films obtained from stacked metallic precursors for solar cell application. *Mater. Lett.* **2015**, *160*, 468–471. [CrossRef]
42. Ming, H.; Zhu, C.; Qin, X.; Jabar, B.; Chen, T.; Zhang, J.; Xin, H.; Li, D.; Zhang, J. Improving the thermoelectric performance of Cu2SnSe3: Via regulating micro-and electronic structures. *Nanoscale* **2021**, *13*, 4233–4240. [CrossRef] [PubMed]

*Article*

# Synthesis and Characterization of Al- and SnO$_2$-Doped ZnO Thermoelectric Thin Films

Giovanna Latronico [1,*], Saurabh Singh [2], Paolo Mele [1,*], Abdalla Darwish [3], Sergey Sarkisov [4], Sian Wei Pan [5], Yukihiro Kawamura [5], Chihiro Sekine [5], Takahiro Baba [6,7], Takao Mori [6,7], Tsunehiro Takeuchi [2], Ataru Ichinose [8] and Simeon Wilson [3]

[1] College of Engineering, Shibaura Institute of Technology, 307 Fukasaku, Minuma-ku, Saitama 337-8570, Saitama Prefecture, Japan

[2] Toyota Technological Institute, Tempaku Ward, Hisakata 2-12-1, Nagoya 468-8511, Aichi Prefecture, Japan; saurabhsingh@toyota-ti.ac.jp (S.S.); t_takeuchi@toyota-ti.ac.jp (T.T.)

[3] Physics and Engineering Department, Dillard University, 2601 Gentilly Blvd, New Orleans, LA 70122, USA; adarwish@dillard.edu (A.D.); Swilson@dillard.edu (S.W.)

[4] SSS Optical Technologies, 515 Sparkman Drive, Huntsville, AL 35816, USA; Mazillo123@yahoo.com

[5] Muroran Institute of Technology, 27-1 Mizumotocho, Muroran 050-8585, Hokkaido Prefecture, Japan; 21043057@mmm.muroran-it.ac.jp (S.W.P.); y_kawamura@mmm.muroran-it.ac.jp (Y.K.); sekine@mmm.muroran-it.ac.jp (C.S.)

[6] International Center for Materials Nanoarchitectonics (WPI-MANA), National Institute for Materials Science (NIMS), Namiki 1-1, Tsukuba 305-0044, Ibaraki Prefecture, Japan; BABA.Takahiro@nims.go.jp (T.B.); MORI.Takao@nims.go.jp (T.M.)

[7] Graduate School of Pure and Applied Sciences, University of Tsukuba, 1-1-1 Tennodai, Tsukuba 305-8577, Ibaraki Prefecture, Japan

[8] Central Research Institute of Electric Power Industry (CRIEPI), 2-6-1 Nagasaka, Yokosuka 240-0196, Kanagawa Prefecture, Japan; ai@criepi.denken.or.jp

[*] Correspondence: na19103@shibaura-it.ac.jp (G.L.); pmele@shibaura-it.ac.jp (P.M.)

**Citation:** Latronico, G.; Singh, S.; Mele, P.; Darwish, A.; Sarkisov, S.; Pan, S.W.; Kawamura, Y.; Sekine, C.; Baba, T.; Mori, T.; et al. Synthesis and Characterization of Al- and SnO$_2$-Doped ZnO Thermoelectric Thin Films. *Materials* **2021**, *14*, 6929. https://doi.org/10.3390/ma14226929

Academic Editor: Andres Sotelo

Received: 8 October 2021
Accepted: 10 November 2021
Published: 16 November 2021



**Abstract:** The effect of SnO$_2$ addition (0, 1, 2, 4 wt.%) on thermoelectric properties of *c*-axis oriented Al-doped ZnO thin films (AZO) fabricated by pulsed laser deposition on silica and Al$_2$O$_3$ substrates was investigated. The best thermoelectric performance was obtained on the AZO + 2% SnO$_2$ thin film grown on silica, with a power factor (*PF*) of 211.8 μW/m·K$^2$ at 573 K and a room-temperature (300 K) thermal conductivity of 8.56 W/m·K. *PF* was of the same order of magnitude as the value reported for typical AZO bulk material at the same measurement conditions (340 μW/m·K$^2$) while thermal conductivity $\kappa$ was reduced about four times.

**Keywords:** thermoelectricity; Seebeck coefficient; thermal conductivity; thin film; oxides

## 1. Introduction

The increasing worldwide demand for energy and the resultant depletion of fossil fuels have brought new challenges for the scientific community [1]. One of the major issues is to develop high-efficiency devices for capturing energy from abundant natural sources such as solar, wind and geothermal energy. Another surplus, but mostly unused, source of energy is wasted heat. There are huge waste heat sources in our environments covering a wide range of temperatures (300~1200 K): industrial processes, domestic stoves and radiators, electrical lighting, pipelines, electrical substations, subway networks, automotive exhaust tubes, but also geothermal heat, body heat, and so on: about 66% of the annual world energy consumption is lost as waste heat, and the loss corresponds to the stellar amount of $3 \cdot 10^{20}$ J per year, just considering the past 10 years [2,3]. A highly promising method for energy recovery from such heat sources is the utilization of thermoelectric (TE) materials that can convert various types of waste heat flow into electricity. The possibility of conversion of heat in electricity was discovered in 1821 by Thomas Seebeck [4]: a junction

73

of two metals generates the thermoelectric voltage $\Delta V$ when a temperature gradient $\Delta T$ is created across it. In mathematical form it can be expressed as:

$$\Delta V = S \cdot \Delta T \tag{1}$$

where $S$ is the material-dependent Seebeck coefficient. The performance of any TE material is quantified by the figure of merit ($ZT$):

$$ZT = \frac{\sigma S^2}{\left(\kappa_{el} + \kappa_{ph}\right)} T \tag{2}$$

where $\sigma$ is the electrical conductivity, $S$ the Seebeck coefficient, $T$ the absolute temperature, $\kappa_{el}$ the electrical thermal conductivity and $\kappa_{ph}$ the phononic thermal conductivity being $\kappa = \kappa_{el} + \kappa_{ph}$ the total thermal conductivity.

$ZT$ is related to the efficiency $\eta$ of heat/electrical energy conversion by the relation:

$$\eta = \frac{\Delta T}{T_H} \left( \frac{\sqrt{1 + ZT_{ave}} - 1}{\sqrt{1 + ZT_{ave}} + \frac{T_C}{T_H}} \right) \tag{3}$$

State-of-the-art TE materials for heat conversion can operate at $T = 300{\sim}1200$ K, with $ZT = 0.1{\sim}2.6$, corresponding to $\eta = 1{\sim}20\%$. Therefore, after two centuries, TE devices are seldom utilized in daily life. TE modules have obtained brilliant success, however, in niche applications, for example for powering the space probe Cassini [5], while high-conversion modules to harvest waste heat of car engines remained at the state of prototypes [6]. $ZT$ and $\eta$ must be strongly improved to make TE power competitive with the common thermodynamic cycles based on fossil fuel burning, solar conversion and nuclear power plants. The improvement of $ZT$ can be obtained by enhancing $\sigma$ and/or $S$, or the product $\sigma S^2$ which is called "power factor" (*PF*), and/or by decreasing the total thermal conductivity. However, $\kappa_{el} = LT\sigma$, with $L = 2.44 \times 10^{-8}$ W·$\Omega$·K$^{-2}$ is the Lorentz number while $\kappa_{ph} = 1/3\, Cv\Lambda$ where $C$ is the heat capacity, $v$ is the speed of phonons, $\Lambda$ is the phonon mean free path. It is thus necessary to decouple $\sigma$ and $\kappa_{el}$ and work on the depression of $\kappa_{ph}$.

Since the discovery of the Seebeck effect, the benefit of TE harvesting has been understood and a wealth of TE materials has been discovered. After the discovery of BiTe with $ZT = 0.5$ [7] in 1954, the $ZT$ of TE materials did not improve for a very long time. In 1993, a seminal paper [8] theoretically predicted the drastic depression of $\kappa_{ph}$ due to reduction of the mean free path of phonons ($\Lambda$) to a few nanometers, which resulted in substantial improvement of $ZT$. This concept was firstly validated in Bi$_2$Te$_3$/Sb$_2$Te$_3$ multilayer thin films incorporating natural nano-sized precipitates, showing the extremely small $\kappa$ value (0.22 W·m$^{-1}$ K$^{-1}$) and a huge value of $ZT = 2.5$ at T = 300 K [9].

These results were due to the scattering of phonons by nano-defects spontaneously formed inside the thin films. However, TE harvesters based on BiTe and related materials cannot be produced on a large scale because they contain rare (Te) and poisonous (Sb, Bi, Pb) elements, and require high-cost processing. Though the last ten years a great progress has been made with $ZT = 2.6$ for SnSe single crystals [10] and 3.1 for polycrystalline SnSe [11], there is still room for finding even more efficient TE materials in a wide T region.

In particular, the attention of researchers has been focused on the development of stable, environmentally benign, abundant, and cost-effective TE materials based on oxides. In the 1990s, extensive research in the area of bulk oxides focused on the enhancement of the TE performance using atomic substitutions and improved grain connection [12–15]. To date, the best TE performance of the currently available oxide materials is $ZT = 0.64$ for *n*-type Zn$_{0.96}$Al$_{0.02}$Ga$_{0.02}$O [13] and 0.74 for *p*-type Ca$_{2.5}$Tb$_{0.5}$Co$_4$O$_9$ [14] at 1000 K. The $ZT$ of oxides is not yet up to the level of the best conventional TE materials and needs to be drastically improved to be acceptable for practical applications. The bulk oxides also have the main disadvantages of requiring a long time for sintering and fabrication of the *n*- and *p*- elements and their assembly in modular shape, and mechanical fragility. All these drawbacks can be overcome by using oxide thin films. Thin-films have significant advantages, such as low dimensionality, rapid fabrication, control of strain at the interface with substrates, and the possibility to insert artificial nano-defects to improve the phonon scattering. Thermoelectric thin films are also attractive for their applicative potential for energy harvesting to power Internet of Things (IoT) sensors [15,16].

Recently, our group studied oxide-thin TE films and obtained encouraging preliminary results. First, epitaxial thin films of 2% Al-doped ZnO (AZO) were fabricated by pulsed laser deposition (PLD) on several single crystal ($SrTiO_3$, $Al_2O_3$) and amorphous (silica) substrates. Regardless of a particular substrate, the films always showed higher values of $ZT$ in comparison with the corresponding bulk AZO: for example, at $T = 600$ K, $ZT_{AZO\text{-}on\text{-}STO} = 0.03$ while $ZT_{bulk} = 0.014$ [17,18]. The superior performance of films is due to their lower thermal conductivity: $\kappa_{AZO\text{-}on\text{-}STO}$ (300 K) = 6.5 W/m·K [16,17] while $\kappa_{bulk}$ (300 K) = 34 W/m·K. In these series, the grain boundaries can be considered as natural nanodefects for an enhanced scattering of phonons and consequent decrease of $\kappa$. As a demonstration of this effect, the film on fused silica, showing additional grain boundaries at the seed layer on the substrate, had even lower thermal conductivity: $\kappa_{silica}$ (300 K) = 4.89 W/m·K and larger $ZT$: $ZT_{silica}$ (600 K) = 0.045 [18].

The insertion of artificial nanodefects has been subsequently considered with the purpose of further reduction of $\kappa$ and enhancing $ZT$. Several approaches have been tried by our group: insertion of hydroquinone nanolayers in AZO films prepared by atomic layer deposition (ALD): $\kappa_{ALD}$ (300 K) = 3.56 W/m·K [19]; addition of polymethylmethacrylate (PMMA) particles to AZO films prepared by multi-beam multi-target matrix-assisted PLD (MBMT/MAPLE-PLD): $\kappa_{MAPLE}$ (300 K) = 5.9 W/m·K and $ZT_{MAPLE}$ (600 K) = 0.0061 [20]; formation of nanopores in AZO films prepared by mist-chemical vapor deposition (mist-CVD): $\kappa_{porous}$ (300 K) = 0.60 W/m·K and $ZT_{porous}$ (300 K) = 0.057 [21]; dispersion of $Al_2O_3$ nanoparticulate in AZO films prepared by surface-modified target PLD: $\kappa_{nanoAl2O3}$ (300 K) = 3.98 W/m·K and $ZT_{nanoAl2O3}$ (600 K) = 0.0007 [22]. Several groups worldwide have recently obtained excellent results with different kinds of dopant, like the $\kappa$ (300 K) = 1.19 W/m·K and $ZT$ (300 K) = 0.1 of magnetron sputtered AZO films by Loureiro et al. [23], $\kappa$ (323 K) = 3.37 W/m·K and $ZT$ (423 K) = 0.052 of CNT-added evaporated porous AZO films by Liu et al. [24], the $\kappa$ (300 K) = 1.1 W/m·K and $ZT$ (300 K) = 0.042 of amorphous $ZnO_xN_y$ PLD films by Hirose et al. [25], the $\kappa$ (300 K) = 1.8 W/m·K and $ZT$ (383 K) = 0.019 of dual-doped AlGaZnO PLD films by Nguyen, et al. [26], just to cite the most impressive. All these successful examples highlight the promise of nanostructured doped ZnO films for future energy-harvesting applications.

In this work, we have focused on enhancing the thermoelectric performance of Al-doped ZnO films by the addition of a controlled amount of $SnO_2$ as a secondary dopant. AZO has the same structure of hexagonal ZnO: wurtzite, hexagonal, space group **$P6_3mc$**, cell parameters $a = b = 0.3289$ nm and $c = 0.5307$ nm, while $SnO_2$ is cassiterite, tetragonal, space group **$P4_2/mnm$** with cell parameters $a = b = 0.4832$ nm and $c = 0.3243$ nm. Since the mismatch between the $a$-axis of AZO and the $c$-axis of $SnO_2$ is 1.4%, we forecasted the growth of quasi-epitaxial nanostructures (nanoparticles) of $SnO_2$ with $c_{SnO2}//a_{AZO}$, dispersed in the AZO matrix.

The motivation of this work is to prove the presence of $SnO_2$ nanoparticulates in the AZO matrix and verify the optimal content of $SnO_2$ for the TE properties of the nanocomposite Al- and $SnO_2$-doped (AZO + $SnO_2$) thin films.

## 2. Materials and Methods

Two sets of four thin films were prepared using various compositions and substrate materials. All samples were deposited both on amorphous silica and $c$-axis oriented alumina by pulsed laser deposition (PLD) technique focusing a Nd:YAG (266 nm, 10 Hz) laser on dense pellets. Four commercial targets (Kurt Lesker Ltd., Jefferson Hills, PA, USA) used for film making had different contents of $SnO_2$ (0, 1, 2 and 4 wt.%) dispersed in the Al-ZnO (AZO) main phase where the Al content was kept as 2 wt.%. Prior to the deposition, the substrates were cleaned at 773 K for 2 h, then glued by conductive silver glue to a rectangular Inconel plate, which was put in direct thermal contact with a cartridge heater and finally inserted into the vacuum chamber. All films were grown with an energy density of about 4.2 J/cm$^2$ for 60 min at 573 K, in the atmosphere of 20 mtorr of oxygen. The samples deposited on silica are named AZO_xS and the ones deposited on alumina AZO_xA, with x = 0, 1, 2 and 4 referring to the concentration (wt.%) of the SnO$_2$ in the respective targets.

Electrical conductivity ($\sigma$) and Seebeck coefficient (S) were measured by means of the four-probe method using a ZEM-3 (ULVAC Advance-Riko, Yokohama, Kanagawa, Japan) apparatus under partial He pressure to assure thermal transport between the heater and the sample. Measurements with ZEM-3 were conducted performing three cycles between 353 and 573 K and measuring both heating and cooling processes. The $\sigma$ and S curves reported in Section 3.2 for each sample were those obtained from the last cooling cycle, considered as the most stable and reliable. The uncertainties of the ZEM-3 measurements were ±4% for the electrical conductivity and ±3% for the Seebeck coefficient.

For comparison and validation of data, some selected samples were also characterized by a custom-made apparatus whose working principle and measurement protocol are reported elsewhere [27,28]. The error in the thermoelectric coefficients measurements is ±5% in the case of σ, and ±7% for S.

The thickness of the thin films (in the range 295–585 nm, as summarized in Table 1) was measured using a Dektak 6M Stylus Profiler (Bruker, Billerica, MA, USA) and a Filmetrics Profilm 3D (KLA, Milpitas, CA, USA).

The X-ray diffraction analysis was conducted using a Bragg-Brentano powder diffractometer (Smart Lab 3, Rigaku Corporation, Tokyo, Japan) using the Cu K$_\alpha$ radiation in the 10–100° angular range with an angular step of 0.02° (power settings: 40 mA, 40 kV).

The TEM images and EDS analysis were conducted using a JEM-2100F (JEOL, Akishima, Tokyo, Japan) with EDS spectroscopy. The STEM beam size was set to 0.5 nm.

The picosecond time-domain thermo-reflectance (TD-TR) technique using a customized system PicoTR (PicoTherm, Netzsch Japan KK, Tsukuba, Japan) was utilized to measure the thermal conductivity of the samples at room temperature (T = 300 K) in the cross-plane direction.

Measurement of Hall coefficient at room temperature was performed using a standard four-terminal method with a commercial PPMS instrument (Quantum Design, San Diego, CA, USA) at an AC current of 5 mA.

## 3. Results and Discussion

### 3.1. Structural and Morphological Characterization

The XRD spectra of the thin films fabricated on amorphous silica and single-crystal c-axis oriented Al$_2$O$_3$ are presented in Figure 1a,b, respectively.

On both substrates, the preferred orientation of AZO films was along the c-axis since only (00l) reflection appeared. The presence of SnO$_2$ in the films of both series was undetectable by X-rays even for the highest amount (x = 4).

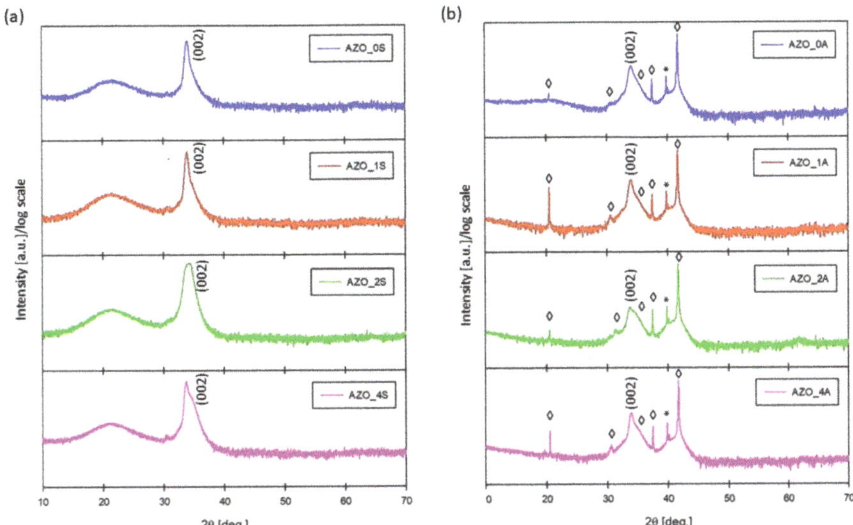

**Figure 1.** X-ray diffraction (XRD) spectra of Al-doped ZnO (AZO) films deposited (**a**) on silica (series AZO_xS); (**b**) on Al₂O₃ (series AZO_xA). *(hkl)* reflections of AZO [18] are indicated, while ◊ labels the peaks of Al₂O₃ substrate and asterisks * the unattributed peaks.

The cross-sectional TEM images (magnification 150,000, Figures 2a and 3a of two representative samples deposited on fused silica without (AZO_0S) and with the addition of SnO₂ (2 wt.%, AZO_2S) reveal a typical columnar growth along the c-axis of the films as often reported in the literature for AZO PLD thin films [16,17,19,21]. The addition of SnO₂ (AZO_2S) did not affect the morphology of the films. The higher magnification images (300,000×, Figures 2b and 3b for AZO_0S and AZO_2S, respectively) show that the columns were well connected without the presence of pores or misoriented grains so that the density of the films could be considered close to the theoretical value.

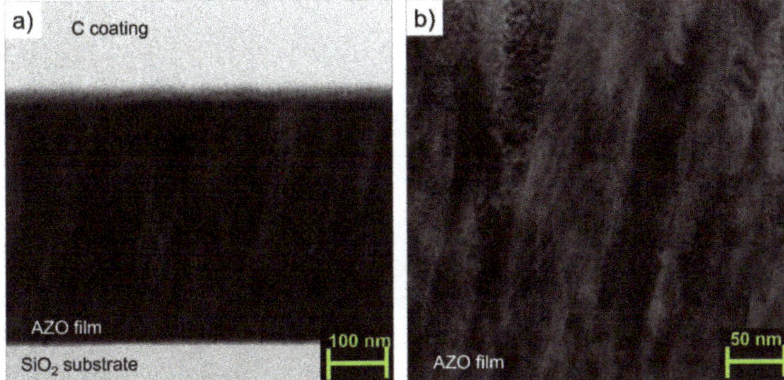


**Figure 2.** (**a**) Cross-sectional transmission electron microscope (TEM) image (magnification 150,000) of the AZO/silica thin film AZO_0S; (**b**) double magnification (300,000×) taken in the central zone of Figure 2a.

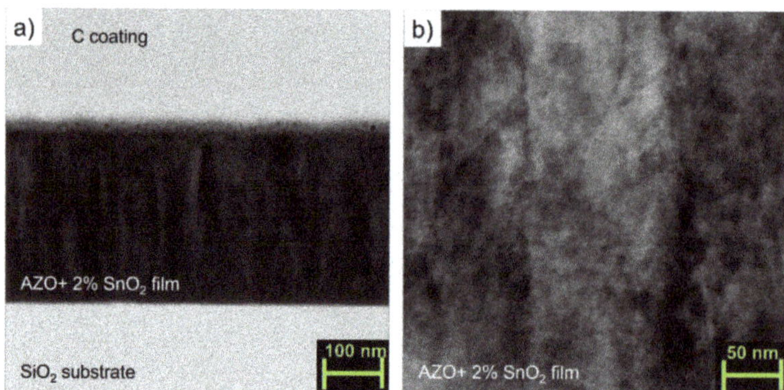



**Figure 3.** (**a**) Cross-sectional TEM image (magnification 150,000) of the AZO/silica thin film AZO_2S; (**b**) double magnification (300,000×) taken in the central zone of Figure 3a.

The TEM-EDS mapping performed on the sample AZO_2S (Figure 4) reveals a uniform distribution not only of the elements Al, Zn, O, but also of Sn. Contrary to the predictions based on the 1.4% lattice mismatch, Sn did not form nanoscale aggregates in the AZO matrix, but presumably entered as substitutional or interstitial atoms in the AZO elementary cell. According with TEM-EDS analysis, the atomic percentage of Sn in the film prepared with 2 wt.% $SnO_2$-added AZO target is 0.49%. The atom % of oxygen is 33.91%. Since it is not possible to separate the contribution of oxygen from $SnO_2$ and from ZnO, we can only compare the percentage of Sn in the target and in the film. The percentage of Sn in $SnO_2$ is 78.77%, which in 2 wt.% $SnO_2$ corresponds to 1.58%. This means that in the film the amount of Sn is about 1/3 than in the target. Si is not considered in the discussion being part of the substrate only.

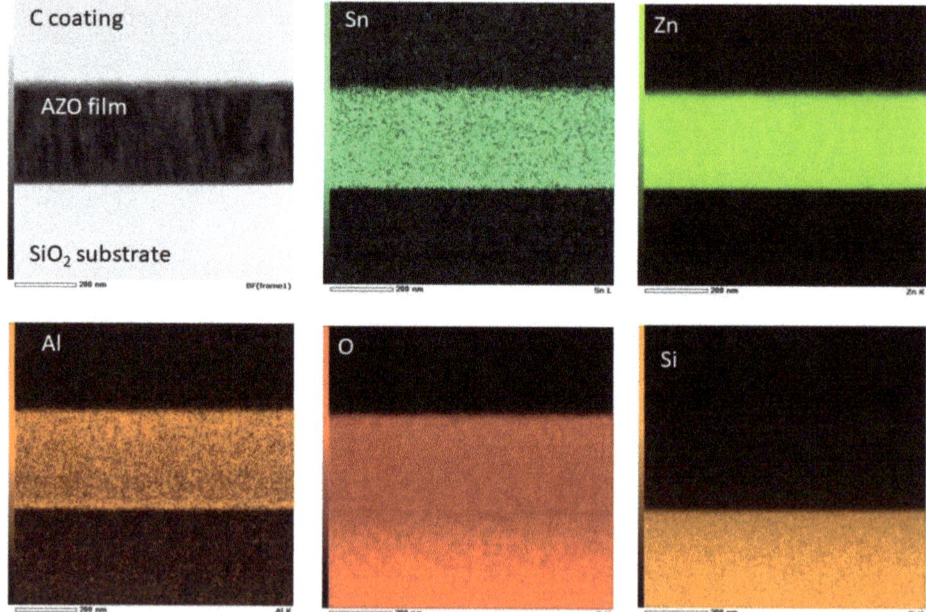



**Figure 4.** TEM-EDS (energy-dispersive spectroscopy) elemental maps of the AZO + $SnO_2$ 2 wt.% silica thin film (AZO_2S) for the atoms Sn, Zn, Al, O and Si.

### 3.2. Transport and Thermoelectric Characterization

Room temperature Hall measurement was taken to find the carriers' concentration $n_H$ and mobility $\mu_H$ values for the AZO_xS and AZO_xA films. The results are in Figure 5, together with room temperature resistivity values from the ZEM-3 analysis.

The resistivity at room temperature has a clear ascending trend with the increase of the content of $SnO_2$ for both substrates. The films on $Al_2O_3$ had half the resistivity of those on silica. The range of $n_H$ was almost the same for two substrates, and of the order of $10^{20}$ cm$^{-3}$, as expected for a typical semiconductor such as AZO. The maximum number of carriers occurred at x = 1 on silica and at x = 2 on $Al_2O_3$. The carrier mobility had the maximum at x = 2 on both substrates but was about 20 times larger for AZO_2A than for AZO_2S. AZO_0S had greater mobility than AZO_0A and this could be explained by the larger misfit dislocations generated on $Al_2O_3$, due to a lattice misfit of 15% [18,29]. Surprisingly, by increasing the content of $SnO_2$ doping, there was a reversal trend and the films grown on $Al_2O_3$ presented higher mobility, with the maximum value for AZO_2A-20 times larger than for AZO_0S.

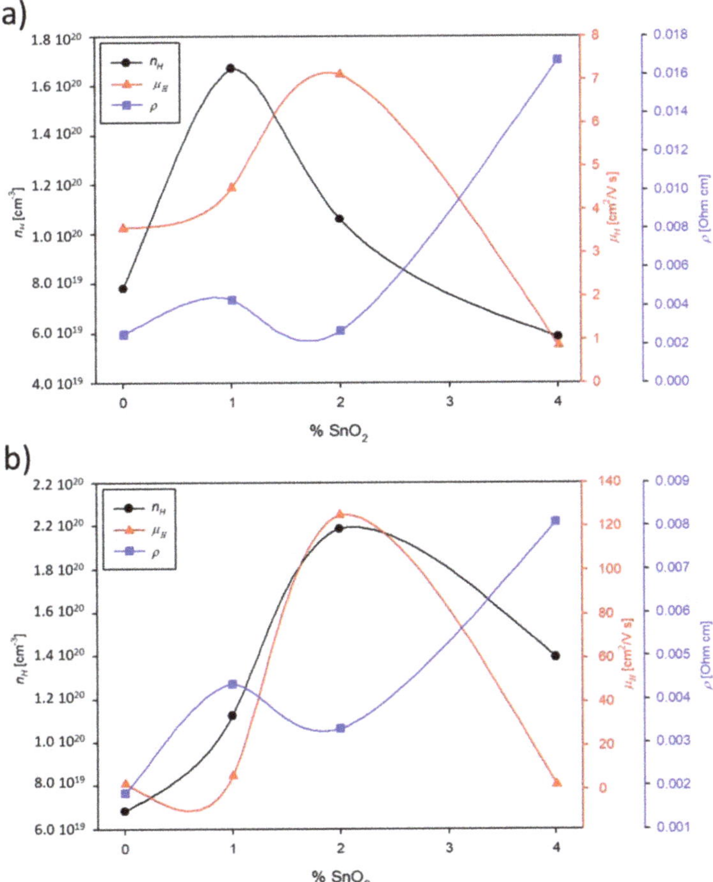

**Figure 5.** Hall effect measurements of the number of carriers and their mobility for films deposited (**a**) on silica (series AZO_xS) and (**b**) $Al_2O_3$ (series AZO_xA) plotted versus the % content of $SnO_2$ (x). Note that reported is the absolute value of $n_H$ with the aim of simplifying the reading. The resistivity of both series was measured at the lowest temperature (348 K) using the ZEM-3 apparatus.

The electrical conductivity ($\sigma$) for the AZO films is presented in Figure 6 as a function of temperature in the range 300–600 K. Figure 6 shows that $\sigma$ increases with rising temperature, confirming the semiconductor behaviour of the samples. On both substrates, the film grown without additional $SnO_2$ had the largest $\sigma$ followed by the film with x = 2.

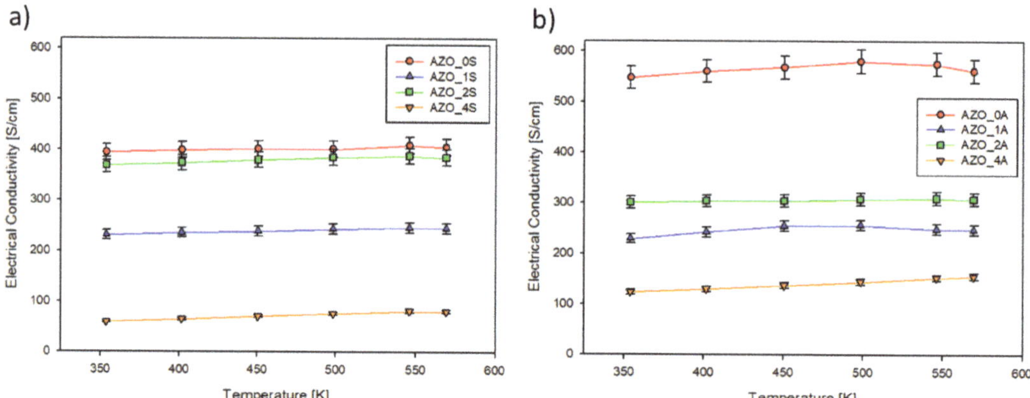

**Figure 6.** Electrical conductivity plotted vs. temperature of thin films deposited on (**a**) silica (series AZO_xS); (**b**) on $Al_2O_3$ (series AZO_xA).

The temperature dependence (300–600 K) of Seebeck coefficient ($S$) of the AZO films is plotted in Figure 7. S had a negative sign, confirming the *n*-type conductivity of AZO. Figure 7 shows that $S$ decreases with the rising temperature, a reversal trend with respect to $\sigma$ as commonly expected for semiconductor materials. On both substrates, the pure AZO film had the lowest $S$ while the sample fabricated with a target containing 4% of $SnO_2$ presents the largest one.

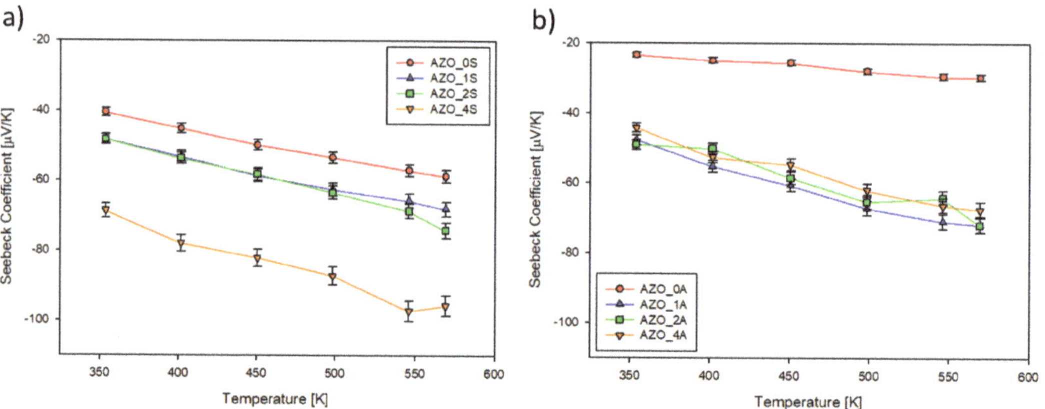

**Figure 7.** Seebeck coefficient ($S$) plotted vs. temperature of thin films deposited on (**a**) silica (series AZO_xS); (**b**) on $Al_2O_3$ (series AZO_xA).

The Seebeck coefficient can be expressed as [30]:

$$S = \left(\frac{8\pi^2 k_B^2}{3eh^2}\right) m^* T \left(\frac{\pi}{3n}\right)^{2/3} \tag{4}$$

where $k_B$ is Boltzmann's constant, $h$ is Planck's constant, $T$ is the absolute temperature, $m^*$ is the effective mass and $n$ is the carrier concentration. The effective mass is also expressed as $m^* = \alpha\, m_e$, where $m_e$ is the mass of the electron and $\alpha$ is a positive rational number.

The room temperature (300 K) "Pisarenko plots" ($S$ versus $n_H$) for both series of films are presented in Figure 8. According to the graphs, $m^*$ was (0.3–0.6) $m_e$ and (0.2–0.7) $m_e$ for the films grown on silica and alumina, respectively. The fractional me was consistent with the parabolic band approximation and relatively simple band structure reported for ZnO [31].

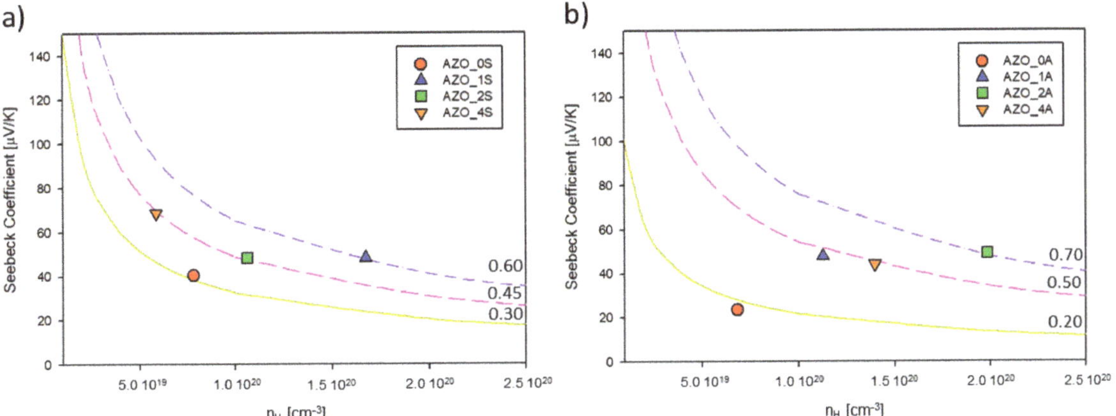

**Figure 8.** Pisarenko plot at 300 K for thin films deposited on (**a**) silica (series AZO_xS) and (**b**) on $Al_2O_3$ (series AZO_xA).

The calculated power factor (*PF*) as a function of temperature is shown in Figure 9. The maximum power factor value was found for sample AZO_2S at 573K as 211.78 mW/m·K$^2$. This value was of the same order of magnitude as that measured for bulk AZO at the same conditions ($-340$ mW/m·K$^2$) [32] and was also comparable with the values reported for AZO thin films [17,18,23–26].

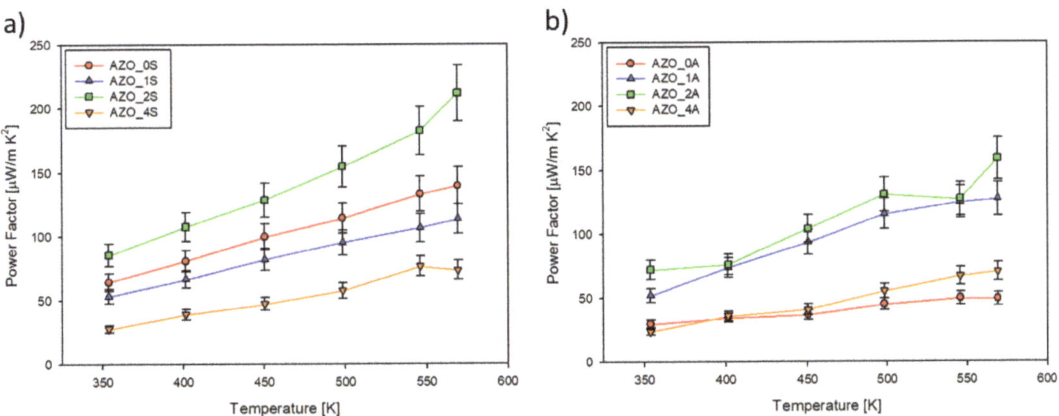

**Figure 9.** Power factor $PF = \sigma S^2$ plotted vs. temperature for thin films deposited on (**a**) silica (series AZO_xS) and (**b**) on $Al_2O_3$ (series AZO_xA).

The *PF*s of all films at 353 K and 573 K are summarized in Figure 10, making clear that on both substrates, 2% of $SnO_2$ represents the optimal dopant concentration.

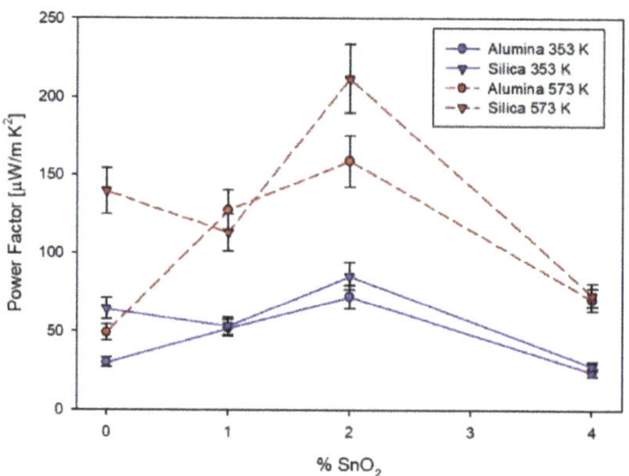



**Figure 10.** The trend of power factor with the content of $SnO_2$ for the two series of films on silica and alumina.

For the films AZO_0S and AZO_2S deposited on silica the measurements were repeated using two different apparatuses to verify the accuracy of the reproducibility of data. The comparison between the pair sets of data for $\sigma$, $S$ and $PF$ is reported in Appendix C. The transport and thermoelectric properties of the films are summarized in Table 1.

**Table 1.** Thickness, electrical and thermoelectric properties of the AZO films deposited on silica (series AZO_xS) and on $Al_2O_3$ (series AZO_xA) for different concentrations x (%) of $SnO_2$ dopant.

| Sample | Thickness [nm] | $n_H$ [cm$^{-3}$] | $\mu_H$ [cm$^2$/V·s] | $\sigma$ [S/cm] | | S [µV/K] | | PF [µW/m·K$^2$] | | $k$ [W/m·K] |
|---|---|---|---|---|---|---|---|---|---|---|
| | | (300 K) | (300 K) | (353 K) | (573K) | (353 K) | (573K) | (353 K) | (573K) | (300 K) |
| AZO_0S | 595 | $-7.812 \cdot 10^{19}$ | 3.571 | 394.7 | 405.3 | −40.41 | −58.66 | 64.47 | 139.5 | 8.88 |
| AZO_1S | 505 | $-1.671 \cdot 10^{20}$ | 4.503 | 231.2 | 245.0 | −48.07 | −68.02 | 53.43 | 113.4 | - |
| AZO_2S | 300 | $-1.061 \cdot 10^{20}$ | 7.119 | 369.2 | 385.0 | −48.13 | −74.17 | 85.54 | 211.8 | 8.56 |
| AZO_4S | 295 | $-5.865 \cdot 10^{19}$ | 0.8606 | 59.63 | 80.23 | −68.57 | −95.73 | 28.03 | 70.70 | - |
| AZO_0A | 535 | $-6.832 \cdot 10^{19}$ | 2.576 | 547.8 | 561.4 | −23.38 | −29.61 | 29.94 | 49.21 | - |
| AZO_1A | 498 | $-1.124 \cdot 10^{20}$ | 6.345 | 229.8 | 247.7 | −47.66 | −71.72 | 52.19 | 127.4 | - |
| AZO_2A | 535 | $-1.986 \cdot 10^{20}$ | 125.1 | 301.4 | 307.6 | −49.00 | −71.87 | 72.38 | 158.9 | - |
| AZO_4A | 480 | $-1.394 \cdot 10^{20}$ | 2.143 | 123.7 | 155.4 | −44.06 | −67.45 | 24.01 | 70.70 | - |

The room temperature (300 K) thermal conductivity ($\kappa$) of the sample with the highest $PF$ (AZO_2S) and of a reference sample from the same batch (AZO_0S) was evaluated by the TD-TR method [32–35] according to the details described in Appendix A. The value of $\kappa$ was 8.88 and 8.56 W/m·K for samples AZO_0S and AZO_2S, respectively. These values were in the same range as those reported for pure AZO films deposited by PLD [16,17,21]. Therefore, the four-fold decrease of $\kappa$ with respect to bulk [32] must be attributed to the phonon scattering of grain boundaries, and not to the addition of $SnO_2$. The calculation of $ZT$ ($T = 353$ K) using Equation (2) gave the following result: $ZT$ (AZO_0S) = 0.003 and $ZT$ (AZO_2S) = 0.004, of the same order as for AZO thin film deposited by PLD on different substrates. Conservative estimates of $ZT'$ ($T' = 573$ K) were $ZT'$ (AZO_0S) = 0.009 and $ZT'$ (AZO_2S) = 0.014. Since for oxide thin films $\kappa$ is expected to decrease with $T$ [33], $ZT$ was calculated using a conservative approach from the $PF$ values at 353 K or 573 K, $\kappa$ at 300 K, and $T = 353$ K or 573 K.

These values of $ZT$ did not surpass the reported performance of bulk AZO at the same conditions and were also of the same or lower level as for other AZO films prepared by PLD. This indicated that $SnO_2$ could not be considered as an optimal dopant for AZO thin films for thermoelectric applications. On the other hand, it could be that the fine dispersion of $SnO_2$ into AZO matrix enhanced the photoelectric response of AZO, as reported for $SnO_2/ZnO$ hierarchical nanostructures prepared by the electrospinning method [34], or could be used to fabricate compact gas sensors, as reported for $ZnO$–$SnO_2$ nanofibers [35].

## 4. Summary

In summary, we fabricated using pulsed laser deposition (PLD) two series of thermo-electric thin films of Al-doped ZnO (AZO) doped with various concentrations of $SnO_2$ (x = 0, 1, 2, 4 wt.%) on the substrates made of fused silica and $Al_2O_3$ (001) single crystals. The goal was to enhance the thermoelectric performance. All the films were c-axis oriented. The films deposited on silica showed the highest values of Seebeck coefficient (*S*) and power factor *(PF)* in comparison with the films on $Al_2O_3$ with the same content of $SnO_2$. The AZO film on silica with x = 2 showed the best performance, with $\sigma$ = 385.0 S/cm, *S* = –74.17 mV/K, and *PF* = 211.8 $\mu W/m \cdot K^2$ at the maximum operating temperature (573 K). The room temperature (300 K) thermal conductivity of this sample was evaluated as $\kappa$ = 8.56 W/m·K, with a calculated figure of merit *ZT* (353 K) = 0.003 and a projected value of *ZT* (573 K) = 0.014, comparable with *ZT* of the nanostructured PLD AZO films grown on several substrates.

**Author Contributions:** Conceptualization, P.M. and G.L.; Data curation, G.L. and S.S. (Saurabh Singh); formal analysis, G.L.; funding acquisition, T.M. and A.D.; investigation, G.L., S.S. (Saurabh Singh), S.W.P., Y.K., S.W. and T.B.; methodology, G.L. and P.M.; resources, P.M., T.M., T.T., A.I. and A.D.; supervision, P.M.; visualization, G.L.; writing—original draft preparation, G.L. and P.M.; writing—review and editing, G.L., S.S. (Saurabh Singh), S.W.P., Y.K., C.S., T.M., S.S. (Sergey Sarkisov), A.D. and P.M. All authors have read and agreed to the published version of the manuscript.

**Funding:** A.D. appreciates the financial support from US Air Force Office of Scientific Research Grant FA9550-18-1-0364 AFOSR, Army Research Office Grant No W911-NF-19-1-0451, and the partial support from Dillard University Minority Health and Health Disparity Research Center MHHDRC. T.B and T.M. acknowledge support from JST Mirai Program JPMJMI19A1 and JSPS KAKENHI JP19H00833.

**Institutional Review Board Statement:** Not applicable.

**Informed Consent Statement:** Not applicable.

**Data Availability Statement:** The data presented in this research study are available in this article.

**Acknowledgments:** P.M. and G.L. acknowledge the endorsement of the International Research Center for Green Electronics (IRCGE), Shibaura Institute of Technology. P.M. and G.L. appreciate Shrikant Saini (Kyushu Institute of Technology, Kitakyushu, Japan) and Marco Antonio de la Torre Hidalgo (University Castilla-La Mancha, Ciudad Real, Spain) for the insightful discussions. P.M. acknowledges Masami Yamashita of Filmetrics, KLA corporation, for her support on part of the thickness measurements. P.M. and G.L. thank Cristina Artini (University of Genova, Genova, Italy) for the support on XRD data interpretation.

**Conflicts of Interest:** The authors declare no conflict of interest. The funders had no role in the design of the study; in the collection, analyses, or interpretation of data; in the writing of the manuscript, or in the decision to publish the results.

## Appendix A

The picosecond time-domain thermo-reflectance (TD-TR) was utilized to measure the thermal conductivity of two samples in the cross-plane direction using the front-heat front-detect (FF) configuration. Prior to the measurements, a 100 nm layer of Pt was sputtered on the surfaces of the thin films (see Figure A1). As a reference sample, we used 100 nm-thick Pt film deposited on a silica substrate. Rear-heat front-detect (RF) configuration was used for the reference sample (see Figure A1b). Figure A3 shows the thermo-reflectance signal obtained from the reference sample. We determined 1946 $[J/(s^{0.5} \, m^2 \, K)]$ as the thermal effusivity of the silica substrate by curve fitting. (see Figure A1b).

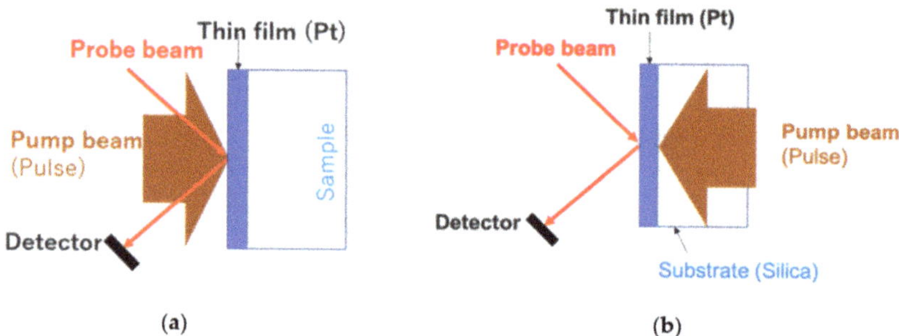



**Figure A1.** Diagram of thermo-reflectance measurement with (**a**) front-detect (FF) configuration and (**b**) rear-heat front-detect (RF) configuration.

After depositing the Pt layer on the surface, samples consisted of three layers (see Figure A2).

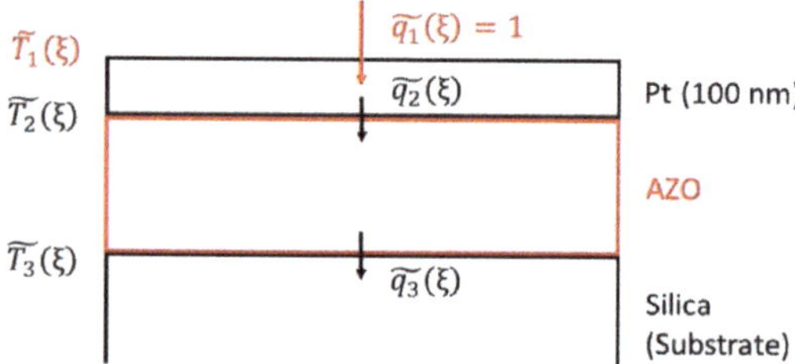



**Figure A2.** Diagram of the applied model.

The thermal conductivities of the thin films were determined by the picosecond time-domain thermo-reflectance (TD-TR) technique, using the original focused beam system [36–38].

For the analysis, the mirror image method was used [39]. In the mirror image method, the temperature history of FF configuration could be described as:

$$T_f(t) = \frac{1}{b_f \sqrt{\beta t}} \left( 1 + 2 \sum_{n=1}^{\infty} fl^n \exp\left( -n^2 \frac{\varnothing_f}{t} \right) \right) \tag{A1}$$

$$\varnothing_f = \frac{d_f{}^2}{ff_f} = \left( \frac{C_f}{b_f} \right)^2 \tag{A2}$$

$$fl = \frac{b_f - b_s}{b_f + b_s} \tag{A3}$$

where $\varnothing_f$ is the heat diffusion time (in Pt layer), $fl$ is the dimensionless parameter, $b_f$ is the thermal effusivity of the Pt layer, $b_s$ is the thermal effusivity of the sample, $d_f$ is the thickness of the Pt layer and $ff_f$ is the thermal diffusivity of the Pt layer.

We already knew the value of $C_f$, thus we could determine the thermal effusivity of the sample $b_s$. We could also determine the thermal conductivity of the samples $\kappa_s$ using the following equation:

$$\kappa_s = \frac{b_s{}^2}{c_s \rho_s} \tag{A4}$$

where $c_s$ and $\rho_s$ are the specific heat capacity and density of the AZO layer, respectively.

The aforementioned model assumed that the sample consisted of two layers and the second layer (AZO) was semi-infinite. This model could be derived from the relationship between the temperature and heat density of the sample layers using quadrupole matrix expression [39]:

$$\begin{bmatrix} \tilde{q}_2(\zeta) \\ \tilde{T}_2(\zeta) \end{bmatrix} = \begin{bmatrix} \cosh\sqrt{\tau_1\tilde{\zeta}} & -b_f\sqrt{\tilde{\zeta}}\sinh\sqrt{\tau_1\tilde{\zeta}} \\ -\frac{1}{b_1\sqrt{\tilde{\zeta}}}\sinh\sqrt{\tau_1\tilde{\zeta}} & \cosh\sqrt{\tau_1\tilde{\zeta}} \end{bmatrix}$$
$$\begin{bmatrix} \tilde{q}_1(\zeta) \\ \tilde{T}_1(\zeta) \end{bmatrix} \tilde{T}_2(\zeta) = \frac{1}{b_2\sqrt{\zeta}}\tilde{q}_2(\zeta) \tag{A5}$$

However, we found that the effect of the third layer (silica substrate) was significant, so we modified the model in order to account for the substrate. In the three-layered sample, quadrupole matrices were expressed as:

$$\begin{bmatrix} \tilde{q}_3(\zeta) \\ \tilde{T}_3(\zeta) \end{bmatrix} = \begin{bmatrix} \cosh\sqrt{\tau_2\tilde{\zeta}} & -b_2\sqrt{\tilde{\zeta}}\sinh\sqrt{\tau_2\tilde{\zeta}} \\ -\frac{1}{b_2\sqrt{\tilde{\zeta}}}\sinh\sqrt{\tau_2\tilde{\zeta}} & \cosh\sqrt{\tau_2\tilde{\zeta}} \end{bmatrix} \begin{bmatrix} \cosh\sqrt{\tau_1\tilde{\zeta}} & -b_1\sqrt{\tilde{\zeta}}\sinh\sqrt{\tau_1\tilde{\zeta}} \\ -\frac{1}{b_1\sqrt{\tilde{\zeta}}}\sinh\sqrt{\tau_1\tilde{\zeta}} & \cosh\sqrt{\tau_1\tilde{\zeta}} \end{bmatrix} \begin{bmatrix} \tilde{q}_1(\zeta) \\ \tilde{T}_1(\zeta) \end{bmatrix}$$
$$\tilde{T}_3(\zeta) = \frac{1}{b_3\sqrt{\zeta}}\tilde{q}_3 \tag{A6}$$
$$(\zeta)\tau_2 = \left(\frac{C_2}{b_2}\right)^2$$

where $C_2$ is the heat capacity of the AZO layer, $b_3$ is the thermal effusivity of the silica substrate. We could derive the temperature response of the three-layered sample by solving this equation. This time we used the thermal diffusivity of the silica substrate determined using the reference sample ($b_3$ = 1946 J/(s$^{0.5}$ m$^2$ K).

Figure A3 shows the thermo-reflectance signal obtained from the samples. Table A1 shows the values determined by curve fitting.

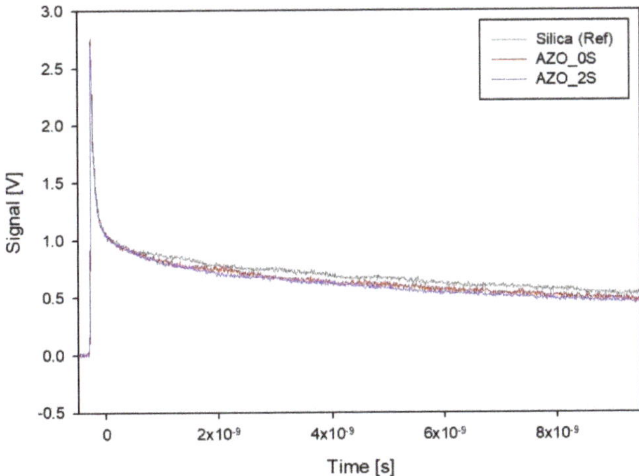


**Figure A3.** Thermo-reflectance signals of the silica reference and AZO samples.

**Table A1.** Thermal properties of samples AZO_0S and AZO_2S.

| Sample | Heat Diffusion Time $\varnothing_f$[s] | $fl$ | Thermal Effusivity $b_s$ [J/(s$^{0.5}$· m$^2$· K)] | Thermal Conductivity $\kappa_s$ [W/(m· K)] |
|---|---|---|---|---|
| AZO_0S | $5.19 \cdot 10^{-10}$ | 0.541 | 3738 | 8.88 |
| AZO_2S | $5.24 \cdot 10^{-10}$ | 0.529 | 3854 | 8.56 |

## Appendix B

The electrical conductivity $\sigma$ is plotted versus the SnO$_2$ concentration x in Figure B1.

The undoped sample had the highest $\sigma$, and x = 2 was the optimal dopant concentration.

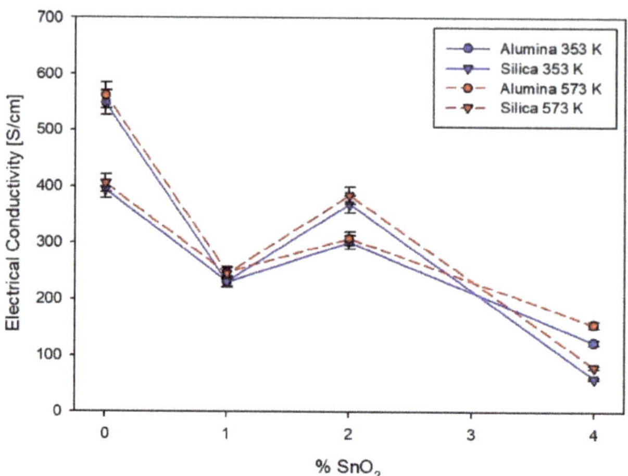


**Figure B1.** Electrical conductivity plotted versus the concentration of $SnO_2$ for the two series of films on silica and alumina.

The Seebeck coefficient $S$ is plotted versus the concentration x of $SnO_2$ in Figure B2. The undoped sample had the lowest $S$ while the sample with x = 4 had the largest $S$.

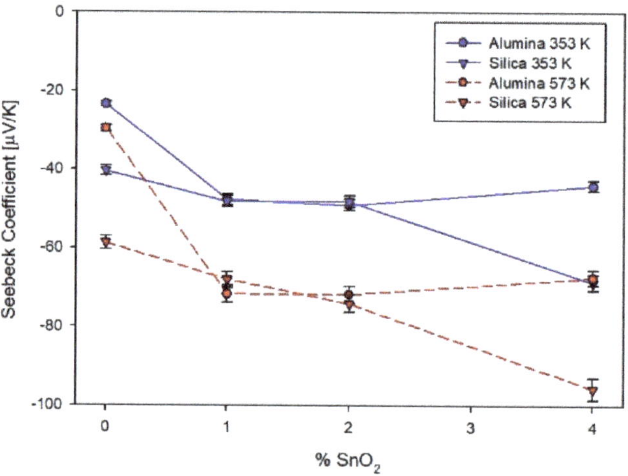


**Figure B2.** Seebeck coefficient plotted versus the concentration of $SnO_2$ for the two series of films on silica and alumina.

## Appendix C

The comparison of $\sigma$, $S$ and $PF$ curves obtained for the samples AZO_0S and AZO_2S is reported in the Figures C1–C3, respectively. Except for the case of s, the pair of curves are comparable within the experimental errors.

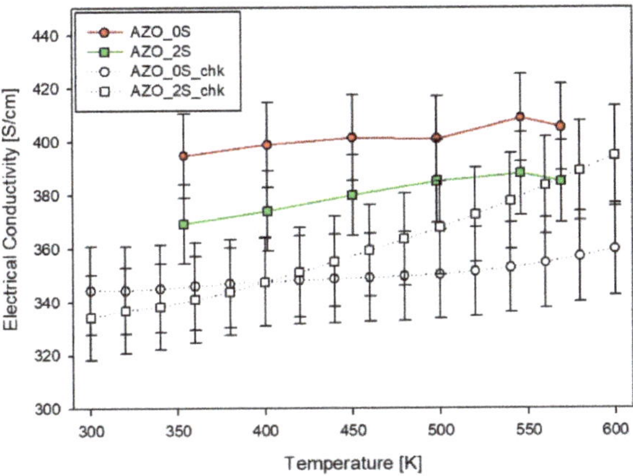

**Figure C1.** Electrical conductivity plotted vs. temperature of AZO_0S and AZO_2S obtained with two different setups. Filled symbols are the same as those reported in the plots in the main text, while the open symbols "chk" were obtained with another device.

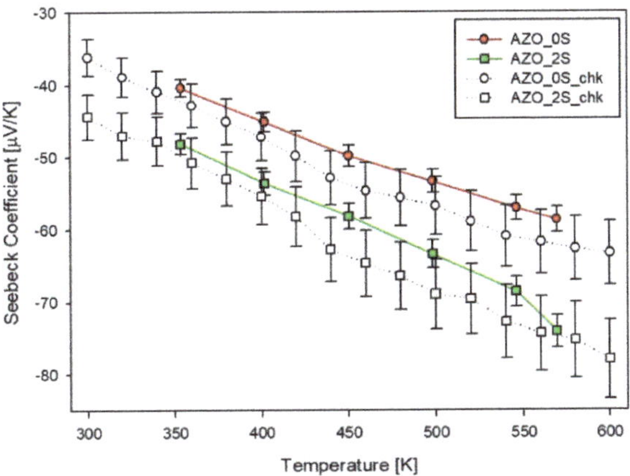

**Figure C2.** Seebeck coefficient plotted vs. temperature of AZO_0S and AZO_2S obtained with two different setups. Filled symbols are the same as those reported in the plots in the main text, while the open symbols "chk" were obtained with another device.

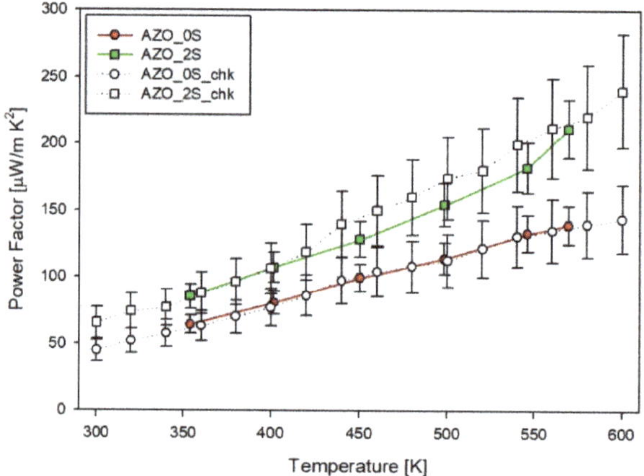

**Figure C3.** Power factor plotted vs. temperature of AZO_0S and AZO_2S calculated from experimental data obtained with two different setups. Filled symbols are the same as those reported in the plots in the main text, while the open symbols "chk" were obtained with another device.

## References


1. Chu, S.; Majumdar, A. Opportunities and challenges for a sustainable energy future. *Nature* **2012**, *488*, 294. [CrossRef] [PubMed]
2. BP Statistical Review of World Energy. Available online: https://www.bp.com/en/global/corporate/energy-economics/statistical-review-of-world-energy.html (accessed on 20 August 2021).
3. Annual Energy Flow Charts by Lawrence Livermore Nat. Lab. Available online: https://flowcharts.llnl.gov/commodities/energy (accessed on 20 August 2021).
4. Seebeck, T.J. Ueber die magnetische polarisation der metalle and erze durch temperatur-differenz. *Ann. Phys.* **1821**, *82*, 133–160.
5. Mireles, V.; Stultz, J.W. Radioisotope Thermoelectric Generator Waste Heat System for the Cassini Propulsion Module. *J. Aerosp.* **1994**, *103*, 548.
6. Crane, D.; LaGrandeur, J.; Jovovic, V.; Ranalli, M.; Adldinger, M.; Poliquin, E.; Dean, J.; Kossakovski, D.; Mazar, B.; Maranville, C. TEG On-Vehicle Performance and Model Validation and What It Means for Further TEG Development. *J. Electron. Mater.* **2013**, *42*, 1582–1591. [CrossRef]
7. Goldsmid, H.J.; Douglas, R.W. The use of semiconductors in thermoelectric refrigeration. *Br. J. Appl. Phys.* **1954**, *5*, 386. [CrossRef]
8. Hicks, L.D.; Dresselhaus, M.S. Effect of quantum-well structures on the thermoelectric figure of merit. *Phys. Rev. B* **1993**, *47*, 12727–12731. [CrossRef] [PubMed]
9. Venkatasubramanian, R.; Siivola, E.; Colpritts, T. Thin-film thermoelectric devices with high room-temperature figures of merit. *Nature* **2001**, *413*, 597–602. [CrossRef]
10. Zhao, L.D.; Lo, S.H.; Zhang, Y.; Sun, H.; Tan, G.; Uher, C.; Wolverton, C.; Dravid, V.P.; Kanatzidis, M.G. Ultralow thermal conductivity and high thermoelectric figure of merit in SnSe crystals. *Nature* **2014**, *508*, 373–377. [CrossRef] [PubMed]
11. Zhou, C.; Lee, Y.K.; Yu, Y.; Byun, S.; Luo, Z.Z.; Lee, H.; Ge, B.; Lee, Y.L.; Chen, X.; Lee, J.Y.; et al. Polycrystalline SnSe with a thermoelectric figure of merit greater than the single crystal. *Nat. Mater.* **2021**, *20*, 1378–1384. [CrossRef]
12. Fergus, J.W. Oxide materials for high temperature thermoelectric energy conversion. *J. Eur. Ceram. Soc.* **2012**, *32*, 525–540. [CrossRef]
13. Ohtaki, M.; Araki, K.; Yamamoto, K. High Thermoelectric Performance of Dually Doped ZnO Ceramics. *J. Electron. Mater.* **2009**, *38*, 1234. [CrossRef]
14. Saini, S.; Yaddanapudi, H.; Tian, K.; Yin, Y.; Magginetti, D.; Tiwari, A. Terbium Ion Doping in $Ca_3Co_4O_9$: A Step towards High-Performance Thermoelectric Materials. *Sci. Rep.* **2017**, *7*, 44621. [CrossRef]
15. Petsagkourakis, I.; Tybrandt, K.; Crispin, X.; Ohkubo, I.; Satoh, N.; Mori, T. Thermoelectric materials and applications for energy harvesting power generation. *Sci. Technol. Adv. Mater.* **2018**, *19*, 836–862. [CrossRef] [PubMed]
16. Kishore, R.A.; Priya, S. A Review on Low-Grade Thermal Energy Harvesting: Materials, Methods and Devices. *Materials* **2018**, *11*, 1433. [CrossRef]
17. Mele, P.; Saini, S.; Honda, H.; Matsumoto, K.; Miyazaki, K.; Hagino, H.; Ichinose, A. Effect of substrate on thermoelectric properties of Al-doped ZnO thin films. *Appl. Phys. Lett.* **2013**, *102*, 253903. [CrossRef]
18. Saini, S.; Mele, P.; Honda, H.; Henry, D.J.; Hopkins, P.E.; Molina-Luna, L.; Matsumoto, K.; Miyazaki, K.; Ichinose, A. Enhanced thermoelectric performance of Al-doped ZnO thin films on amorphous substrate. *Jpn. J. Appl. Phys.* **2014**, *53*, 060306. [CrossRef]

19. Tynell, T.; Giri, A.; Gaskins, J.; Hopkins, P.E.; Mele, P.; Miyazaki, K.; Karppinen, M. Efficiently suppressed thermal conductivity in ZnO thin films via periodic introduction of organic layers. *J. Mater. Chem. A* **2014**, *2*, 12150–12152. [CrossRef]

20. Darwish, A.M.; Muhammad, A.; Sarkisov, S.S.; Mele, P.; Saini, S.; Liu, J.; Shiomi, J. Thermoelectric properties of Al-doped ZnO composite films with polymer nanoparticles prepared by pulsed laser deposition. *Compos. Part B Eng.* **2019**, *167*, 406–410. [CrossRef]

21. Saini, S.; Mele, P.; Oyake, T.; Shiomi, J.; Niemelä, J.P.; Karppinen, M.; Miyazaki, K.; Li, B.; Kawaharamura, T.; Ichinose, A.; et al. Porosity-tuned thermal conductivity in thermoelectric Al-doped ZnO thin films grown by mist-chemical vapor deposition. *Thin Solid Film.* **2019**, *685*, 180–185. [CrossRef]

22. Mele, P.; Saini, S.; Mori, T.; Kakefuda, Y.; Adam, M.I.; Singh, S.J. Thermoelectric properties of Al2O3-doped AZO thin films. article in preparation.

23. Loureiro, J.; Neves, N.; Barros, R.; Mateus, T.; Santos, R.; Filonovich, S.; Reparaz, S.; Sotomayor-Torres, C.M.; Wyczik, F.; Divay, L.; et al. Transparent aluminium zinc oxide thin films with enhanced thermoelectric properties. *J. Mater. Chem. A* **2014**, *2*, 6649–6655. [CrossRef]

24. Liu, S.; Lan, M.; Li, G.; Piao, Y.; Ahmoum, H.; Wang, Q. Breaking the tradeoff among thermoelectric parameters by multicomposite of porosity and CNT in AZO films. *Energy* **2021**, *225*, 120320. [CrossRef]

25. Hirose, Y.; Tsuchii, M.; Shigematsu, K.; Kakefuda, K.; Mori, T.; Hasegawa, T. Thermoelectric properties of amorphous $ZnO_xN_y$ thin films at room temperature. *Appl. Phys. Lett.* **2019**, *114*, 193903. [CrossRef]

26. Nguyen, N.H.T.; Nguyen, T.H.; Liu, Y.-R.; Aminzare, M.; Pham, A.T.T.; Cho, S.; Wong, D.P.; Chen, K.-H.; Seetawan, T.; Pham, N.K.; et al. Thermoelectric Properties of Indium and Gallium Dually Doped ZnO Thin Films. *ACS Appl. Mater. Interfaces* **2016**, *8*, 33916–33923. [CrossRef]

27. Singh, S.; Pandey, S.K. Fabrication of Simple Apparatus for Resistivity Measurement in High-Temperature Range 300–620 K. *IEEE Trans. Instrum. Meas.* **2018**, *67*, 2169–2176. [CrossRef]

28. Byeon, D.; Sobota, R.; Singh, S.; Ghodke, S.; Choi, S.; Kubo, N.; Adachi, M.; Yamamoto, Y.; Matsunami, M.; Takeuchi, T. Long-Term Stability of the Colossal Seebeck Effect in Metallic $Cu_2Se$. *J. Electron. Mater.* **2020**, *49*, 2855–2861. [CrossRef]

29. Novotny, M.; Cızek, J.; Kuzel, R.; Bulır, J.; Lancok, J.; Connolly, J.; McCarthy, E.; Krishnamurthy, S.; Mosnier, J.-P.; Anwand, W.; et al. Structural characterization of ZnO thin films grown on various substrates by pulsed laser deposition. *J. Phys. D Appl. Phys.* **2012**, *45*, 225101. [CrossRef]

30. Snyder, C.J.; Toberer, E.S. Complex thermoelectric materials. *Nat. Mater.* **2008**, *7*, 101. [CrossRef] [PubMed]

31. Yaakob, M.K.; Hussin, N.H.; Taib, M.F.M.; Kudin, T.I.T.; Hassan, O.H.; Ali, A.M.M.; Yahya, M.Z.A. First Principles LDA + U Calculations for ZnO Materials. *Integr. Ferroelectr.* **2014**, *155*, 15–22. [CrossRef]

32. Mele, P.; Kamei, H.; Yasumune, H.; Matsumoto, K.; Miyazaki, K. Development of Thermoelectric Module Based on Dense $Ca_3Co_4O_9$ and $Zn_{0.98}Al_{0.02}O$ Legs. *Met. Mater. Int.* **2014**, *20*, 389–397. [CrossRef]

33. Nolas, G.S.; Goldsmid, H.J. *Thermal Conductivity: Theory, Properties and Applications*; Kluwer Academic/Plenum: New York, NY, USA, 2004; p. 114.

34. Zhu, L.; Hong, M.; Ho, G.W. Hierarchical Assembly of $SnO_2$/ZnO Nanostructures for Enhanced Photocatalytic Performance. *Sci. Rep.* **2015**, *5*, 11609. [CrossRef] [PubMed]

35. Lu, Z.; Zhou, Q.; Wang, C.; Wei, Z.; Xu, L.; Gui, Y. Electrospun $ZnO–SnO_2$ Composite Nanofibers and Enhanced Sensing Properties to SF6 Decomposition Byproduct $H_2S$. *Front. Chem.* **2018**, *6*, 540. [CrossRef] [PubMed]

36. Kakefuda, Y.; Yubuta, K.; Shishido, T.; Yoshikawa, A.; Okada, S.; Ogino, H.; Kawamoto, N.; Baba, T.; Mori, T. Thermal conductivity of $PrRh_{4.8}B_2$, a layered boride compound. *APL Mater.* **2017**, *5*, 126103. [CrossRef]

37. Piotrowski, M.; Franco, M.; Sousa, V.; Rodrigues, J.; Deepak, F.L.; Kakefuda, Y.; Baba, T.; Kawamoto, N.; Owens-Baird, B.; Alpuim, P.; et al. Probing of thermal transport in 50 nm Thick PbTe nanocrystal films by time-domain thermoreflectance. *J. Phys. Chem. C* **2018**, *122*, 27127–27134. [CrossRef]

38. Daou, R.; Pawula, F.; Lebedev, O.; Berthebaud, D.; Hebert, S.; Maignan, A.; Kakefuda, Y.; Mori, T. Anisotropic thermal transport in magnetic intercalates $Fe_xTiS_2$. *Phys. Rev. B* **2019**, *99*, 085422.

39. Baba, T. Analysis of One-dimensional Heat Diffusion after Light Pulse Heating by the Response Function Method. *Jpn. J. Appl. Phys.* **2009**, *48*, 05EB04. [CrossRef]

MDPI
St. Alban-Anlage 66
4052 Basel
Switzerland
Tel. +41 61 683 77 34
Fax +41 61 302 89 18
www.mdpi.com

*Materials* Editorial Office
E-mail: materials@mdpi.com
www.mdpi.com/journal/materials